CLARENDON ARISTOTLE SERIES

General Editors

J. L. ACKRILL AND LINDSAY JUDSON

ARISTOTLE
PHYSICS

BOOK VIII

Translated
with a Commentary
by

DANIEL W. GRAHAM

CLARENDON PRESS · OXFORD
1999

Oxford University Press, Great Clarendon Street, Oxford OX2 6DP
Oxford New York
Athens Auckland Bangkok Bogotá Buenos Aires Calcutta
Cape Town Chennai Dar es Salaam Delhi Florence Hong Kong Istanbul
Karachi Kuala Lumpur Madrid Melbourne Mexico City Mumbai
Nairobi Paris São Paulo Singapore Taipei Tokyo Toronto Warsaw
and associated companies in
Berlin Ibadan

Oxford is a registered trade mark of Oxford University Press

Published in the United States
by Oxford University Press Inc., New York

British Library Cataloguing in Publication Data
Data available

Library of Congress Cataloging in Publication Data
Data available

ISBN 0-19-824091-0
ISBN 0-19-824092-9 (Pbk.)

1 3 5 7 9 10 8 6 4 2

Typeset by Best-set Typesetter Ltd., Hong Kong
Printed in Great Britain
on acid-free paper by
Bookcraft Ltd.
Midsomer Norton, Somerset

ΑΡΤΕΜΙΔΙ ΑΜΥΜΟΝΙ ΑΛΟΧΩΙ

ACKNOWLEDGEMENTS

I began work in earnest on this project in the 1993–4 academic year, supported by a fellowship from the National Endowment for the Humanities and a professional development leave from Brigham Young University. I also received support during several summers from the College of Humanities at my university. I have profited from discussions of my ideas with colleagues on numerous occasions, including a conference on Aristotle's Philosophical Development held at Boston University in January 1992. The translation received careful scrutiny from John Ackrill, and the commentary from Lindsay Judson, both of whom have been patient and helpful through a series of revisions. To these institutions and individuals I acknowledge my indebtedness. And to my wife Diana and my children Sarah and Joseph I give loving thanks for their moral support.

D. W. G.

Provo, Utah
May 1998

CONTENTS

CONTENTS

INTRODUCTION

Physics VIII can be viewed in the way that Aristotle may himself have viewed it, as the crowning achievement of his theory of nature. The work is difficult to classify by modern standards: it is at once about 'physics' in Aristotle's sense of a science of natural motion; about what we might call rational cosmology, dealing with the causal and theoretical conditions of the cosmos; and about the metaphysical presuppositions of physics and cosmology. In it Aristotle moves from a discussion of motion in the cosmos to the identification of a single source of all motion. And to identify this source of motion is to solve the ultimate riddle of science: to locate the principle that both produces and regulates motion in the universe, the One beside the Many that eluded the early cosmologists.

Before Socrates, natural philosophers typically explained our world by telling how and from what it came into being, and how it functions now. Many of the philosophical questions that naturally arise from such a study remained unanswered, some unasked. If the world came to be at a certain time, what existed before that time? Why did the world come to be just when it did, and why did it come to be at all? Indeed, did the world come to be, or was the order we experience now always there? What is the source of cosmic order? Does that source belong to the natural world, or does it transcend it?

It is here that we see Aristotle's superiority to his predecessors. Where they had simply assumed a primeval motion, Aristotle expressly asks what right we have to believe in such a motion, what its nature is, and what follows from such assumptions. Where they made implicit assumptions, Aristotle makes explicit enquiries. Where they had explained the generation of the world, he explores the presuppositions of the world's existence. For the first time Aristotle exposes the assumptions of philosophical cosmology to critical scrutiny and raises the discussion of cosmology to the level of rigorous philosophical speculation. Aristotle had already laid the foundations of philosophical cosmology in the *De Caelo* (which I presume was mostly written earlier than *Physics* VIII), but there he discusses cosmology from the standpoint of the characteristic motions of natural bodies in the cosmos. The sum total of these motions I shall call 'cosmic motion'; *Physics* VIII is about what we are justified in inferring from the existence of cosmic motion. The answers Aristotle gives are very different from contemporary answers. But in his relentless pursuit of ultimate presuppositions, we can recognize a kindred spirit to that of the modern cosmologist.

Before we can discuss cosmic motion further, we must get clear on the notion of a cosmos. The Presocratic philosophers had generally (following earlier mythological assumptions manifest in Homer and Hesiod—see Stokes 1962–3; KRS, ch. 1) seen our world as an island of order in a vast sea of unstructured matter. This sea was infinite in extent, or at least vaguely 'boundless', constituting the universe (*to pan*). Within the endless expanse, natural processes began a chain of events that produced an ordered world, finite in extent, roughly spherical, having within it an earth at the centre and heavenly bodies surrounding the earth. This world comes to be known by the term *kosmos*, a word implying order and beauty, in the late sixth and early fifth centuries BC (Kerschensteiner 1962). The first generation of Presocratics, the early Ionians, saw the universe as consisting of a single stuff of determinate nature (water, air, fire) or indeterminate nature (the boundless of Anaximander). The original stuff was differentiated and organized into an ordered world by some process involving a circular rotation, whose residual motion is visible in the motion of the heavenly bodies about the earth. The earth was typically viewed as a stationary disc surrounded by a cylindrical or spherical heaven revolving about it. The original stuff is conceived of not as an inert body of matter, but as in some sense intelligent and active, metaphorically as a pilot that 'steers all things' so as to guide the formation of the cosmos.

In the early fifth century Parmenides criticized the natural enquiries of his predecessors, which, he claimed, presupposed the existence of what-is-not; but since the notion of what-is-not is unintelligible, no theory that makes reference to it can be allowed. There can be no coming to be (from what-is-not), no differentiation (from what-is-not), no motion or change (from what-is-not), and the real world must be like a well-rounded sphere. After him the 'pluralists' posited a plurality of realities having the properties that Parmenides attributed to Being, each changeless in nature but able to combine with other beings to form a natural world. Empedocles posited the four elements of earth, water, air, and fire; Anaxagoras an indefinitely large number of stuffs such as air, ether (the shining upper air), earth, flesh, and bone; and the atomists an infinite number of microscopic solid atoms. By positing a domain of changeless elements or atoms, the pluralists could have a world of becoming without having the basic realities themselves come to be. They could thus hope philosophically to defend a belief in nature against the charge that it was incoherent. Armed with a pluralistic ontology, Empedocles described the world as alternating between a cosmos and a sphere of complete mixture of the elements (like Parmenides' sphere), Anaxagoras derived a single expanding cosmos, and the atomists a plurality of cosmoi in an infinite void.

Common to all the Presocratic cosmologists was an assumption that in some sense the phenomena of nature resulted from natural processes. In the place of Zeus the cloud-gatherer or Poseidon the earth-shaker, interactions of natural substances could account for the events in the heavens, on the earth, and under the earth. While early Ionian thinkers tended to endow the original substance with thought and foresight, in explanations of natural events they appealed to physical and mechanical causes. After Parmenides the pluralists tended to separate the forces of nature from the matter: Empedocles personified Love and Strife, the forces of attraction and repulsion, while Anaxagoras posited a cosmic Mind that started the vortex in the primeval chaos. Although Anaxagoras attributed intelligence and foresight to Mind (B12), he did not explain the world in terms of the purposes of Mind. Empedocles made extensive use of chance collocations of elements and compounds in his cosmology (B53, 59, 75, 78, 85, 98, 103, 104; *Phys.* II. 4, 196^{a}17–24). The atomists rejected teleological explanation altogether, appealing instead to purely mechanical interactions of the atoms for the generation of the cosmos and everything in it.

For Plato, the one glimmer of philosophical insight among the early cosmologists was Anaxagoras's assertion that Mind had ordered all things, implying that all things in the world were arranged for the best (*Phd.* 97b–d). If we could actually explain how things were arranged for the best, we would attain a new level of intelligibility in natural philosophy, showing how goodness pervaded the world. In the *Timaeus* Plato expounds a cosmos consisting of a spherical earth surrounded by a spherical heaven; there is nothing outside the heaven to attack and destroy it, and there is no other world than our own. Soul pervades the heaven and causes the orderly motions in it. In a myth-like cosmogony, Plato describes how the demiurge, the cosmic craftsman, fashions the world out of disorderly and recalcitrant materials so that it can exemplify the order and beauty of an ideal archetype. Thus value and order are put into the world by its organizer, and are preserved by the operation of cosmic soul. Plato's conception is expressly designed to combat the godless and valueless mechanism into which scientific explanation had fallen (see *Laws* X). But Plato admits that his account is at best a 'likely tale', for the changing phenomena of the world do not admit of rigorous scientific knowledge (*Tim.* 29b–d).

As an heir to the cosmological tradition and as a student of Plato, Aristotle is sympathetic to both sides. Like the Presocratics, Aristotle believes that a scientific account of the cosmos is both possible and worthwhile, while, like Plato, he believes that mechanical interactions of matter cannot account for the order and beauty of the cosmos. Aristotle's cosmos is influenced by Plato's: he too envisages a unique world consisting of a spherical earth surrounded by a spherical heaven with nothing outside it. And he too sees the need for a world that is teleologically ordered. But he rejects Plato's account of cosmic motion as caused by self-moving soul, as well as the need for a demiurge to organize the cosmos in the first place. Aristotle's cosmos is everlasting, without beginning or end. In that cosmos, explained in the *De Caelo*, there is a region from the moon outward to the circumference of the cosmos in which there is no change or alteration except for motion in place, where everything is composed of an ethereal fifth element that naturally moves in a circle around the earth. Below the moon, the four elements—earth, water, air, and fire, occupying concentric regions (with some intermixture)—move with their characteristic motions, earth and water downward, air and fire upward. These four elements can be transformed into one another, but their overall quantities remain approximately the same. Beneath the moon all things change, come to be, and decay, and no individual thing lasts for ever; above the moon, no perishable thing can exist, and no heavenly body can cease to exist.

In *Physics* II. 1 Aristotle distinguishes between natural and artificial bodies. Unlike artificial bodies, natural bodies have their own source of motion and rest: that is, they move and stop by themselves. Thus earth and water move down, fire and air move up; animals move around their environments, and plants grow; and finally, the heavenly bodies move in circles around the earth. Besides natural motions, there are forced motions, which carry a body contrary to the way it would travel by nature. For instance, if I throw a rock up, it is propelled in a direction contrary to its natural motion. While forced motions are brought about by external agents, natural motions are brought about by the agency of the thing itself.

In *Physics* II. 3 Aristotle also introduces his four-cause theory. There are four causal factors in terms of which all explanations proceed: the material cause or matter out of which a thing is made, the formal cause or structure, the moving (efficient) cause or agency which brings about a change, and the final cause or purpose which the change serves. As an inherent source of change and rest, a nature is already an efficient cause. But Aristotle argues that both matter and form, in different contexts, can be identified with the nature of a thing (ch. 1). In chapter 8 of *Physics* II Aristotle argues that the regularity of natural changes presupposes that natural events in the cosmos are directed towards ends. Thus they are not like chance events, *pace* Empedocles. Aristotle sees nature as a unified system of ends; indeed, he argues that nature belongs to the class of final causes. He makes no reference, however, to any single source of motion in the cosmos; nor does he imply or allude to the existence of an unmoved mover in his main argument. And it remains obscure how the natural motions of the several natural bodies constitute a unified system of ends. In any case, he concludes that nature falls into the class of the final cause.

At this point we might raise a number of questions about the relationship between *Physics* II and *Physics* VIII. Is the latter work the completion of a theory implicit in the former? Is the latter an extension or a modification of the former? Is it then compatible or incompatible with the former? (Von Arnim (1931) and Solmsen (1960: 232–4, 100–2) argue for a development that is inconsistent, Guthrie (1933–4) for one that is consistent.) In point of composition, Book VIII is evidently written later than Books I–VI and the bulk of the *De Caelo*, and is conceived as an independent treatise

(Ross I–II). It seems to build on, and go beyond, the discussion of 'moved movers' of *Physics* VII (see Wardy 1990: 101 and 93 ff.). *Metaphysics Λ*, in turn, seems to presuppose the argument of *Physics* VIII, which then becomes a standard treatment of the causal structure of motion. (References or allusions to the argument of *Phys.* VIII are found in *GC* II. 10, 337^{a}17–22; *MA* 1, 698^{a}7–11; *Met.* Λ7, 1073^{a}5–12; cf. Λ6, 1071^{b}10–11; cf. also *Cael.* I. 7, 275^{b}18–23, which is arguably a later insertion: von Arnim 1931, Easterling 1976.) Scholars interested in the development of Aristotle's thought have drawn on this sequence and on fragments of lost works to produce a scheme of development concerning cosmic motion like this: (i) the early Aristotle had a theory similar to one of Plato's theories, that the heavenly bodies are moved by an internal soul; (ii) later, Aristotle came to see motion as the result of tendencies to motion in matter; (iii) finally, Aristotle arrived at a theory in which the pure actuality of the first unmoved mover caused all motion by providing a transcendent final cause. (Classic developmental accounts are found in Jaeger 1923: 295 ff., von Arnim 1931, Guthrie 1933–4, Ross 94–102, and Ross 1957. The developmental approach is criticized by Cherniss (1944: 581–602), and rebutted by Solmsen (1960: 308 n. 20); cf. Waterlow (1982: 234 n. 20). More recently, comparisons of Aristotle's theory of heavenly motion with Theophrastus's *Metaphysics*, which seems to react to Aristotle's *Metaphysics Λ*, have suggested to some scholars a date in Aristotle's middle period for the doctrine of the unmoved mover: D. Frede 1971, Devereux 1988, Most 1988, Laks and Most 1993. Graham 1996 focuses on considerations based on the theory of motion itself.) While evidence for stage (i) is suspect, one can still profitably ask about the possibility of a development from (ii) to (iii), and, if one sees evidence for such a change, locate the decisive philosophical move in *Physics* VIII.

According to *Physics* VIII, all motions, even natural motions of the elements, require an external mover. If every motion does indeed presuppose an external mover, we must then seek a further cause, and then a further, until we reach a first mover (on pain of an infinite regress). That mover cannot itself be moved, so we must in the end arrive at an unmoved mover. Furthermore, the eternity and regularity of motion in the world require a first unmoved mover as the ultimate cause of all cosmic motion. This ultimate cause can be identified with God, and the causation he exercises can be identified as final causation. Thus physics and cosmology lead to theology. The step to theology—and metaphysics—does not take place in *Physics* VIII but in *Metaphysics* Λ. Nevertheless, the elegant argument leading from physics to theology is found here in *Physics* VIII. It is, to my mind, one of the tightest and most sustained arguments in the Aristotelian corpus. So tightly woven is it that it lends itself to an extended outline—which I have provided in Appendix I. And surely the integration of physics, cosmology, and theology must stand as one of the greatest intellectual feats of Aristotelian theory.

But despite the elegance of the argument and the synthetic power of the conception, Aristotle's theory of cosmic motion developed here raises significant questions for his thought as a whole. Is the theory of motion elaborated in *Physics* VIII compatible with that of Book II? According to Book II, a natural body originates its own motion; according to Book VIII, no body in motion originates its own motion. Indeed, it is precisely the ability of natural bodies to move by themselves that distinguishes their motion from forced motion, caused by an external agent. Can we still distinguish between natural and forced motion on the new theory, and does the new theory say the same as the old? Moreover, the argument of Book VIII seems to interpret the external cause of change as a moving or efficient cause. Can this be reconciled with the ultimate interpretation in *Metaphysics* Λ of the first unmoved mover as a final cause? What indeed would it mean for the unmoved mover, so sublimely aloof from the natural world, to be an efficient cause? Furthermore, at the end of *Physics* VIII Aristotle denies that the unmoved mover is extended, yet locates it at the circumference of the cosmos, on the assumption that its nearness is important to the motion of the outermost sphere of the heaven. What can we make of this combination of physical and non-physical attributes?

Aristotle does a delicate balancing act between the demands of a naturalistic cosmology and the desiderata of an idealistic value theory. Deeply committed to Plato's programme of locating moral values in the universe, Aristotle must somehow build values into a natural world. He will not countenance a cosmic craftsman; nor does he acknowledge a transcendent world of values. He thus remains without an efficient or a formal cause extrinsic to the cosmos. Moreover, matter by itself cannot supply the principle of order for its own organization, so a material cause is ruled out. Yet Aristotle wants his cosmos to be no less structured than a well-wrought artefact. In *Physics* VIII he finds the source of natural, hence orderly motion not in the apparently self-sufficient natures of *Physics* II, but in a first unmoved mover, a being later to be identified with divinity as a transcendent final cause of the cosmos. From the earliest Ionian theorists down to Aristotle, Greek cosmology aimed at accounting for the order of the cosmos within a naturalistic framework of explanation. Aristotle's attempt is the most sophisticated essay in the genre. Dealing with a more rigorous concept of nature and a more refined notion of value, he traces all motion back to a first cause. But ultimately there is a danger that he has substituted one *deus ex machina* for another. In place of the cosmic pilot of the Ionians or the mythical craftsman of Plato, he gives us a transcendent field marshal (*Met. Λ*10, 1075ᵃ11–15). I suspect that in his effort to make nature maximally intelligible, Aristotle has sacrificed the autonomy of nature to a transcendent source of reality that has more in common with the Platonic Forms than he would like to admit. And I believe that his revision of the theory of natural motion (for he must be revising his theory if his argument is to support the conclusion that there is a first unmoved mover of the cosmos) conflicts with his original theory, for reasons I give in the commentary on Chapter 4. If we were to push the claim that every moved body requires an external mover, we would be compelled to erase the distinction between natural and forced motions, and ultimately to treat natural motion as no different in principle from forced motion. Such a rethinking would have the welcome consequence of moving Aristotelian physics in the direction of modern physics; but it would spell disaster for Aristotle's attempt to make Nature sovereign in the universe.

In the end, the treatise seems to raise more questions than it answers. So too will my commentary. But I believe that, all things considered, a brief commentary like the present one is better suited to raising questions than to answering them. We shall see Aristotle at his best and at his worst, forging links between the natural and the divine, seeking total intelligibility in this most orderly and beautiful cosmos, finding connections where none are to be found, and referring all change back to an impossible unity. Ultimately *Physics* VIII serves as a bridge between physics, cosmology, metaphysics, and theology, and its success must be evaluated in light of large bodies of theory and explanation. A reading of this treatise can stimulate us to rethink Aristotle's positions on major issues, and it is hoped that the present study will facilitate such a rethinking.

TRANSLATION

CHAPTER 1

Did motion come into being at some time, without having existed 250^b11 before, and does it perish again in such a way that nothing is in motion? Or is it instead the case that it neither came into being nor perishes, but instead always existed and always will exist; and being deathless and unceasing, is it present in things as if it were a kind of life belonging to everything composed by nature? Indeed, that there is motion is the view of all who discuss nature, since they 15 describe the origin of the world, and their whole study concerns coming to be and perishing—which could not exist if there were no motion. But those who say there are unlimited worlds, some coming to be and some perishing, are saying there is always motion, for 20 episodes of coming to be and perishing necessarily imply motion in turn; whereas those who say there is a single world, or one that does not always exist, also make corresponding assumptions about motion.

Now if it is possible for there to be a time at which nothing is moving, this must happen in one of two ways. Either it will happen as Anaxagoras describes—for he says that when everything was together and had been at rest for an infinite time, Mind imposed 25 motion and separated things—or it will happen as Empedocles describes, with things being in turn in motion and again at rest—in motion when Love makes the one from many or Strife makes many from one, at rest in the intervening times, as he says:

> Thus, inasmuch as they are wont to grow into one from more   30
> and in turn with the one growing apart they become more,
> they are born and they do not enjoy a steadfast life;   **251^a**
> but inasmuch as they never cease continually alternating,
> they are always immovable in a circle.

By the phrase 'inasmuch as they never cease continually alternating' we must understand him to mean 'from this to that'. So we must examine what is the case concerning these matters; for to see 5 the truth will be useful not only for the study of natural philosophy, but for our enquiry into the first principle as well.

Let us begin first from the definitions we have already laid down in the *Physics*. Now we say that motion is the actuality of the
10 movable in so far as it is movable. It is necessary, therefore, that there should be objects which are able to move with each kind of motion. And even apart from the definition of motion, everyone would grant that what moves must be what is able to move with each kind of motion; e.g. it is the alterable that is altered and what
15 is changeable in its place that travels. So there must be something burnable before it is burnt and something able to burn things before it burns them. Thus, these things too must either come to be at some time if they were not previously in existence, or they must always exist.

Furthermore, if each of the movable things came to be, then before the given motion there must have been another change and
20 motion, in which the thing able to be moved or to cause motion itself came to be. To suppose that such things always pre-existed without any motion taking place is absurd even at first glance, and it must become all the more so on closer examination. For supposing that some things are movables and others potential movers, if at some future time there will be some first mover and a corresponding moved thing, while at another time there is no motion, but
25 only rest, it is necessary for this first mover to undergo a previous change. For there was some cause of its being at rest, for rest is the privation of motion. So before the first change there will have been a previous change.

For some things have a single motion, while some have contrary
30 motions; e.g. fire heats but does not cool, but the science of contraries seems to be one and the same. Something similar seems to happen even in the former case: cold heats when it is removed and retires, just as the knower errs on purpose when he applies his
251ᵇ knowledge perversely. But things that are able to act or to be acted on, or to move things and be moved, respectively, are not able to interact under any conditions whatsoever, but only if they are in a certain condition and approach one another. So one thing causes motion, the other is moved whenever they approach each other, and conditions are such that the one is able to cause motion and the
5 other to be moved. Hence, if the motion did not continue always, clearly the objects were not in a condition such that the one was able to be moved and the other able to cause motion, but one of them had to undergo a change. This would necessarily happen even

with relative things; e.g. if what was not double something is now double, one if not both of the things had to change. Therefore, there will be a change previous to the first change.

(Furthermore, how will there be a before and an after if there is 10
no time? Or how will there be time if there is no motion? If, then, time is a number of motion or a kind of motion, and if there is always time, there must also be everlasting motion. But concerning time at least, all with one exception clearly agree: they say that it 15
does not come to be. (And for this reason Democritus in particular shows how it is impossible for everything to have come to be, since time does not come to be. But Plato alone generates time. For he says that it is coeval with the heaven, and that the heaven has come to be.) If, then, it is impossible for time to be or to be thought of apart from the now, and if the now is a kind of mean, since it has a 20
beginning and an end at the same time—for it is the beginning of the future time and the end of the past—then time must always exist. For the end-point of the time taken to be last will belong to the class of nows (for there is nothing to take in the time beside the now), so that since the now is a beginning and an end, there must 25
always be time on both sides of it. But if there is time, evidently there must also be motion, if indeed time is a kind of property of motion.)

The same argument applies to motion being imperishable. For just as the coming to be of motion implied a change previous to the 30
first motion, so the perishing of motion implies a change later than the last motion. For being moved and being able to be moved do not stop at the same time (e.g. being burned and being able to be burned, for it is possible for something to be burnable when it is not burning); nor do being able to move something and moving it. And so what is able to cause perishing will have to perish when it has 252^a
caused the perishing. And what is able to cause this to perish must in turn perish later—for perishing too is a kind of change. If, then, these things are impossible, clearly there is everlasting motion, not motion at one time, rest at another. Indeed, this suggestion amounts to sheer fantasy. 5

And the same goes for saying that things are naturally thus, and that one must accept this as a principle, which is just what Empedocles would seem to mean by saying that it holds of things by necessity that Love and Strife rule and cause motion in turn, and rest in the intermediate time. Perhaps those who posit a single 10

principle of motion, such as Anaxagoras, would say this as well. But surely there is nothing disorderly in things which happen by or according to nature, for nature is a cause of order in everything. But there is no ratio between one infinite and another, yet all order is ratio. Hence, to suppose that things were at rest for an infinite
15 time, then moved at some moment, without there being any distinguishing feature to account for why the change should happen now rather than earlier, and without there being any order, is inconsistent with the action of nature. For what is natural is either uniform and invariable, e.g. fire invariably travels upwards by nature; or if it is not uniform, it varies according to some ratio. In this
20 case it is better to say with Empedocles and those who share his view that the universe is in turn at rest and again in motion. For such a system has at least some kind of order present in it.

Yet someone who holds this view must not just assert that this happens but must say what the cause is—he should not merely posit something or assume an unargued axiom, but should produce
25 an inductive argument or a deductive proof. The things posited by Empedocles are not themselves causes; nor was it the essence of Love and Strife to cause such movement, but it belongs to the former to combine things, to the latter to separate them. If he is going to specify further how they rule in turn, he must explain under what conditions it happens, such as that there is something which unites men, namely Love, and that enemies flee one another
30 ⟨because of Strife⟩. For he supposes that this holds in the whole cosmos because it clearly applies to some cases. The point about ruling in equal times calls for some argument. And, in general, to think that this is a sufficient principle, that something always is or comes to be in this way, is an erroneous assumption. It is to this that Democritus reduces natural explanation, saying that this
35 is how things happened in the past also. He does not think it is requisite to seek a principle of what is always the case. It is right
252^b to say this for some cases, but wrong to say it for all. For instance, a triangle always has its interior angles equal to 180 degrees; but all the same there is some further cause of its being everlasting. However, of first principles which are everlasting there is no further cause.
5   So much for the thesis that there was not, nor ever will be, a time when there was not, or when there will not be, motion.

CHAPTER 2

It is not difficult to refute the positions which conflict with our own. It is chiefly on account of the following considerations that it would seem possible for motion to exist at some time when it had not existed at all: (1) there is no everlasting change. For every change is naturally from something to something, so these contra- 10 ries between which it takes place must be the limit of each change, and there is no infinite motion. (2) We see that it is possible for something to be moved, though it is not moving and does not have in itself any motion, as we see in the case of inanimate things which, though they are not moving either in any part or as a whole but staying at rest, are moved at some time. But they ought to move 15 either always or never if motion does not come to be when it has not existed before. (3) This is especially apparent in the case of animate things. For although sometimes there is no motion in us and we keep still, nevertheless we do move at a certain moment, and a source of motion comes to be in us from within ourselves, even though nothing from outside moves us. This we do not see 20 in the same way in the case of inanimate things, which are always moved by something from outside. But the animal, we say, moves itself. Thus, if indeed an animal is sometimes perfectly still, motion would come to be in a motionless thing from itself and not from outside. But if this can happen in an animal, why cannot the same 25 thing happen in the universe as well? For if it happens in the microcosm, then also in the macrocosm. And if in the cosmos, then also in the infinite, if indeed it is possible for the infinite to move and to be at rest as a whole.

Point (1), that motion towards opposites is not always the same and one in number, is quite correct. Indeed, it is perhaps necessary, 30 if it is possible that the motion of what is one and the same is not always one and the same. I mean, for example, we may wonder whether one and the same sound comes from the same string, or whether it is different each time, even though the string is in a similar condition and moves in a similar way. But whatever the case, nothing prevents some movement from being the same in 35 virtue of being continuous and everlasting. This will become 253^a clearer from what is said later.

Concerning (2), there is nothing absurd about a thing which was

not in motion being moved if an external cause of motion is sometimes present, sometimes not. We must, however, enquire into how
this can happen—I mean in such a way that the same thing is at one
5 time moved by the same moving agent and sometimes not. For this
is precisely the problem being raised, why it is not the case that
some things are always at rest and others always in motion.

Point (3) would seem to present the most difficult problem: how
a motion that was not present before could come into being later,
which is just what happens in the case of animate things. What
10 was at rest before walks later, although nothing from outside, as it
seems, moves it. This description, however, is false. For we always
see some constituent part of the animal moving. The animal itself
is not responsible for the movement of this part, but perhaps the
environment is. It itself, we say, does not cause all its own move-
15 ments, but only its movements in place. Thus nothing prevents us
from saying—and perhaps indeed we must say—that many motions
come to the body as a result of the environment, and of these some
move the thought or the appetite, and that in turn moves the whole
animal. This happens, for instance, in sleep: although there is no
movement of the senses, some kind of internal motion takes place,
20 and the animals wake up again. But these matters too will become
plain from the following discussion.

CHAPTER 3

The starting-point of our enquiry is also that of the problem that
has just been mentioned: why some beings are sometimes in motion and sometimes in turn at rest. Now, necessarily either (1) all
things are always at rest, or (2) all things are always in motion, or
25 (3) some things are in motion and some at rest; and in the last case
either (a) the things moved are always in motion and the things at
rest always at rest, or (b) all things naturally move and rest alike, or
(c) finally there is still a third alternative. For it is possible for some
beings to be always unmoved, some to be always moving, and some
30 to partake of both states. And this is what we must assert. For it
holds the solution to all our problems, and is the conclusion of our
investigation.

Now to claim that (1) all things are at rest, and to defend this
thesis disregarding sense perception, is a case of intellectual

failure—indeed, it calls into question the whole of experience
rather than some part of it, and not only in relation to the natural 35
scientist, but in relation to virtually all the sciences and all judge- 253^b
ments, since they all make use of motion. Further, objections concerning first principles are not part of a science—as e.g. in the theory
of mathematics they do not concern the mathematician, and similarly in all other cases. So the present debate does not concern the 5
natural scientist. For he *assumes* that nature is a source of motion.

To say (2) that everything is in motion is also virtually a falsehood, but it is less at odds with physics than the former view. For we
have explained in the *Physics* that nature is a principle of rest no
less than of motion; nevertheless, motion is constitutive of nature.
And in fact some thinkers maintain not that some things are in 10
motion and some not, but that all things are in motion always,
though this escapes our senses. Although they do not specify what
sort of motion they mean, or whether they mean all kinds, it is not
difficult to reply to them. For increase and decrease cannot go on
continuously, but there must also be an intermediate state. The
explanation is similar to that about the dripping water wearing 15
away stones and plants splitting stones as they grow out through
them. For if so much material was pushed out or carried away by
the drip, it does not follow that half that much was previously
carried away in half the time. But, as with men dragging a ship, so
this many drops move so much material, but a portion of them do
not move such an amount in any given time. What is taken away
is divisible into several parts; however, none of those parts was 20
moved separately, but all were moved together. Plainly, therefore,
it is not necessary that because decrease is infinitely divisible,
something is being removed at every moment, but only that the
whole disappears at some time.

The same goes for alteration of any kind. For that the thing
altered is infinitely divisible does not imply that the act of alteration
itself is also infinitely divisible, but often the alteration happens all 25
at once, as in the case of freezing. Further, whenever someone falls
ill, there must be a time in which he will recover—he cannot change
in an instant, but the change must be into health and not anything
else. So to say that alteration occurs continuously is to be exces-
sively sceptical of obvious facts. For alteration is toward the contrary state; but a stone becomes neither harder nor softer. As for 30
motion in place, it would be surprising if it escaped our notice

whether a stone was travelling downward or was resting on the
earth. Further, earth and each of the other elements necessarily
remain in their proper places, and they are moved from them only
35 by force. Thus, if some of them are in their proper places, it is
necessary that not all things are moving in place.

254ᵃ   That it is impossible (2) for all things to be always in motion or
(1) for all things to be always at rest, one may conclude from these
and other arguments of this sort. But neither is it possible for (3a)
some things always to be at rest and some things to be always in
motion, but for nothing to be at rest at one time and in motion at
5 another time. We must say that it is impossible in the present cases
as in those previously examined, for we see the above-mentioned
changes occurring in the same things; and, in addition, because the
person who disputes these points is in conflict with obvious facts.
For there will be no increase, nor will forced motion occur, unless
something that was previously at rest is moved contrary to nature.
10 Accordingly, this theory does away with coming to be and perishing. But it is virtually the general view that motion is a kind of
coming to be and perishing. For if something changes into this, it
comes to be this or to occupy this, and if it changes from that, it
ceases to be that or to occupy that. Thus it is clear that some things
sometimes move and some sometimes rest.

15   The view that (3b) all things are at one time at rest and at another time in motion we must now relate to earlier arguments. We
must begin again from our recent distinctions, adopting the same
starting-point as before. Either all things are at rest or all are in
motion, or some entities are at rest and some in motion. And if
some things are at rest and some in motion, then necessarily either
20 all things are at one time at rest and at another time in motion, ⟨or
some are always at rest and others always in motion,⟩ or some are
always at rest, some of them always in motion, and some are now at
rest, now in motion. We have in fact already argued that it is not
possible for all things to be at rest, but we shall argue the point
25 again now. For if things are truly like this, as some claim, and what-
is is infinite and motionless, at least this is not at all how it appears
to the senses, but many beings *seem* to move. If, then, there is false
opinion, or opinion in general, there is also motion, and even if
there is only imagination, or even if opinion varies, still there is
motion. For imagination and opinion are considered to be kinds of
motion.

But to pursue this enquiry and to seek an explanation when we 30
are in the position of not needing one is to be a bad judge of what
is better and what is worse, what is plausible and what is implausi-
ble, and what is fundamental and what is not fundamental. Similarly impossible are the views that everything is in motion and that
some things are always in motion and the other things always at
rest. Against all these views one piece of evidence is sufficient: that 35
we *see* some things are at one time in motion and at another time at **254^b**
rest. So it is apparent that the views that all things are at rest and
that all things are in continuous motion are just as impossible as the
view that some things are always in motion while other things are
always at rest. It remains for us to determine whether all things are
able both to move and to be at rest or whether some things are like 5
this while some are always at rest and some are always in motion.
For it is this that we must now prove.

CHAPTER 4

Of things that cause motion and are moved, some cause motion and
are moved incidentally, some intrinsically—incidentally, e.g. when
they move by belonging to the mover or moved, or by some part of
them moving, intrinsically when they move not by belonging to 10
the mover or moved, or by a part of them moving or being moved.
Of intrinsic motions, some are caused by the thing itself, some by
another thing, and some happen by nature, and others happen
by force and contrary to nature. What is moved by itself is moved
by nature, such as every animal. For an animal is moved by itself, 15
and things which have their source of movement in themselves we
say are moved by nature. That is why the animal as a whole moves
itself by nature; however, its body may be moved both by nature
and contrary to nature. For it makes a difference with what sort
of movement it happens to be moved and from what element it is 20
composed. And of those things moved by another, some are moved
by nature and some contrary to nature; contrary to nature e.g.
earthy bodies moving upward and fire downward, and also the
parts of animals are often moved contrary to nature—contrary to
the normal positions and manners of movement. That a moving
object is moved *by* something is most apparent in the case of things 25
moved contrary to nature, since they are clearly moved by another.

The next most apparent case after things moving contrary to nature is that of things that move themselves by nature, such as animals. For what is unclear in this case is not whether they are moved by something, but rather how we ought to distinguish in them what

30 causes motion from what is moved. For just as in ships and things not composed by nature, so in animals, the mover and the moved part seem to be distinct, and it is in this way that the whole thing moves itself.

What is especially problematic is the remaining member of the last-mentioned division. For of things moved by another we posited

35 that some are moved contrary to nature, but some are left in the
255^a opposite class of being moved by nature. It is these cases—e.g. the light and the heavy—that would create the problem: by what are they moved? For these are moved to the opposite places by force, but to their proper places by nature—the light moving up and the heavy down. But by what they are moved is no longer apparent, as

5 it is when they are moved contrary to nature. For it is impossible to say they are moved by themselves. For this is a property of life, and belongs only to animate things; and in that case they would be able to stop the motion themselves—I mean, for example, if a thing causes its own walking, it also causes its own non-walking—so that if it were in the power of fire itself to travel upward, it would clearly

10 be in its power to travel downward as well. And it is unreasonable also to suppose that things move themselves with only one kind of motion if they do move themselves. Furthermore, how is it possible for something that is continuous and uniform to move itself? For in so far as a thing is one and continuous not by contact, it does not suffer change; but it is in so far as it is divided that one part naturally acts, the other is acted upon. None of the things in ques-

15 tion, therefore, moves itself, for they are uniform, nor does any other continuous thing; but what causes motion must in each case be distinguished from what is moved, as we see in the case of an animate thing moving inanimate things.

Nevertheless, it turns out that these things too are always moved *by* something. This would become apparent if we distinguished the causal factors.

20 What has been said can be applied to movers also. Some are able to move things contrary to nature—e.g. the lever is able to move what is heavy not by nature, while some things are able to move things by nature, as e.g. the actually hot is able to move the poten-

tially hot, and similarly in other such cases. And correspondingly,
what is movable by nature is what is potentially of a certain quality
or quantity or place, when it has such a source in itself and not   25
incidentally (for a certain quality and quantity might belong to
the same subject but the one attribute belongs to the other incidentally and not intrinsically). Fire and earth, then, are moved by
something by force whenever they move contrary to nature, but by
nature whenever they move to the actualities which are potentially   30
theirs.

It is because 'potentially' is said in many ways that it is not
apparent by what such things are moved—e.g. fire up and earth
down. There are different senses in which the learner and one
who already has knowledge but is not exercising it are potentially
knowers. Every time the active and the passive powers come together, the potential always becomes actual; e.g. the learner moves 35
from being potentially one thing to being potentially another thing **255ᵇ**
(for the person who possesses knowledge but is not using it is
potentially a knower in a way, but not in the way he was potentially
a knower before he learned). And whenever he is in this latter
condition, if nothing prevents him, he actualizes and uses his
knowledge—otherwise he would be in the contradictory condition
of ignorance. This is how things are for natural phenomena. For   5
the cold is potentially hot, and whenever it changes, no sooner is it
fire than it burns, unless something prevents and impedes it. This is
also how things are in the case of the heavy and the light. For the
light comes to be from the heavy, e.g. air from water—for it was
first potentially light; and once it is light, it will immediately be 10
active, if nothing prevents it. The actuality of the light is to be
somewhere, namely up, and it is being prevented whenever it is in
the contrary place. And this is how things are also for quantity and
quality.

This brings us to the crucial question: just why do the light and
the heavy move to their own place? The explanation is that it is   15
their nature to go somewhere, and this is what it is to be light or
heavy, the one being defined by *up* and the other by *down*. Things
are *potentially* light and heavy in many ways, as we have stated. For
when something is water, it is potentially light in a way, and when
it is air, it is still potentially light, for something may impede it from
being up. But if the impediment is removed, it becomes active and   20
goes ever upward. It is in the same way that a quality changes into

actually being present. For one who knows immediately uses his knowledge, if nothing prevents him. And a quantity expands in size if nothing prevents. One who removes a prop or an obstacle in a 25 way causes motion and in a way does not; e.g. the man who pulls out a column holding something up or removes a stone from a wineskin in the water: he moves the thing incidentally, just as the ball that bounces back was moved not by the wall but by the thrower. Hence it is clear that none of these things moves itself. But 30 they do have a source of motion, not of causing motion or of acting on something, but of being acted on.

If, then, all things that are in motion are moved either by nature or contrary to nature and by force; and if things moved by force and contrary to nature are all moved by something other than themselves; and if things moved by nature, in turn, are moved by something, both those moved by themselves and those not moved by 35 themselves, e.g. the light and the heavy—for they are moved either 256^a by what generated them and made them light or heavy, or by what removes the impediment or obstacle—then all things that are in motion will be moved by something.

CHAPTER 5

And this will happen in one of two ways: what is in motion will be moved either (a) on account not of the mover itself but of some- 5 thing else that moves the mover, or (b) on account of the mover itself—and this mover will act either (i) first after the last motion or (ii) through several intermediate movers, as when the stick moves the stone and is moved by the hand, which is moved by the man, and he is no longer moved by the motion of anything else. We say that both things cause motion, the last as well as the first mover, but 10 rather the first. For that moves the last, but the last does not move the first, and without the first the last will not move anything, but the first will move without the last—for instance the stick will not move anything if it is not moved by the man. If, then, everything that is in motion must be moved by something, and either (a) by something moved by another or (b) not, and if it is moved by 15 another, there must be some first mover that is not moved by another, but if the immediate mover is of the latter kind, there is no need of the former kind of mover; for it is impossible for a series of movers which are themselves moved by another to go on to infinity, for there is no first member of an infinite series. If, then, everything moved is moved by something, and the first mover is moved, but 20 not by another, it must be moved by itself.

Further, this same argument can be made as follows. Every mover moves something by means of something. For the mover moves either (i) by means of itself or (ii) by means of something else—e.g. a man moves something either by himself or by using a stick, and the wind knocks something down either by itself or by means of the stone which it propelled. It is impossible for a mover 25 to move something without moving by itself that by means of which it causes the motion. But if it moves it by itself, there need not be anything else by which it causes motion; if, however, there is something else by which it causes motion, there will exist something which causes motion not by means of something, but by itself, or the series will go on to infinity. Thus, if something moved moves something, the series must stop and not go on to infinity. For if the stick causes motion by being moved by the hand, the hand moves 30 the stick. And if something else causes motion by means of the hand, something else is the mover of the hand as well. Accordingly, whenever (ii) there is a series in which one thing moves another by means of a third thing, there must be (i) some previous thing causing motion by means of itself. If, then, this thing is moved, but nothing else is its mover, it must move itself. So, according to this argument too, either what is moved is immediately moved by **256ᵇ** something that moves itself or the series eventually arrives at such a thing.

In addition to the previous arguments, these same conclusions will result from the following considerations. If everything that is moved, is moved by a moved, either this attribute belongs to things 5 incidentally, so that what is moved causes motion—not, however, because it itself is moved—or it does not belong to them incidentally but intrinsically. In the first case, if the mover is moved incidentally, it was not necessary for it to be moved. But if so, clearly it is possible that at some time no entity is in motion. For the incidental attribute is not necessary but contingent. Thus, if we 10 assume what is possible, nothing impossible will result, although a falsehood may. But for there to be no motion is impossible. For it has been proved earlier that motion must always exist.

And this result is just what we would expect. For there must be

15 three things: the moved, the mover, and the means. The moved
must be moved, but it need not move anything else. The means
must both cause motion and be moved, for it changes along with
the moved at the same time and in the same respect as it (this is
clear in the case of things that cause motion in place, for the movers
and the moveds must be in contact with each other up to a certain
20 point). But the mover, in so far as it is not the means, is unmoved.
When we observe the last moved, which is able to be moved but
does not have its own source of motion, and what is moved, not by
another but by itself, it is reasonable, not to say necessary, to
suppose that there is a third thing which causes motion while being
unmoved. That is why Anaxagoras speaks rightly when he asserts
25 that Mind is impassible and unmixed, since he makes it the source
of motion. For it would cause motion only by being unmoved, and
would rule only by being unmixed.

But further, if the mover is moved not incidentally but necessarily, and unless it were moved it would not move anything, then the
30 mover, in so far as it is moved, must be moved either according to
the same form of movement as the moved, or according to a different one. I mean, for instance, either the heating agent must itself be
heated, the healing agent be healed, and the transporter be transported, or the healing agent would be transported and the transporter increased in size. But the latter is plainly impossible. For one
257^a must carry the distinction to the individual cases, such as if someone teaches a point of geometry, he is taught this same point, or if
he throws, he is thrown according to the same manner of throwing.
Or it may not happen like this, but each thing may be affected by a
different type of change, such as what transports is increased in
5 size, and the thing that causes this increase is altered by another,
and what causes the alteration is moved with another kind of
motion. But this series must stop. For the kinds of motion are
limited. To return to the beginning of the series and say that what
causes the alteration is transported is tantamount to saying straight
off that the transporter is transported and the teacher is taught.
10 (For it is clear that everything moved is moved also by any antecedent mover in the series, and more especially by the earlier of two
movers.) But this, indeed, is impossible. For it turns out that the
teacher is learning—although the one must lack, the other have
knowledge.

Yet what is even more absurd is that it turns out that everything

which is able to move something is able to be moved, if indeed 15
everything moved is moved by something moved. For it will be
movable, just as if someone said that everything that can heal can
be healed, and everything that can build can be built, either immediately or through several intermediaries. By the latter case I mean
e.g. if everything that can cause motion can be moved by another,
not with the kind of motion with which it moves its neighbour, but 20
with another kind—e.g. what can heal can learn; but by ascending
through the series we arrive at the same form of change, as we said
previously. The former alternative is impossible, the latter fantastic. For it is absurd that what causes alteration should necessarily
be able to be increased. It is not, therefore, necessary for what is
moved in every case to be moved by another, which is itself moved; 25
the series will therefore stop. So either what is first moved will be
moved by something at rest, or it will move itself.

But surely, if it should be necessary to examine whether the
self-mover or the thing moved by another were the cause and
principle of motion, everyone would say the former. For what is a
cause in its own right is always prior to what is itself a cause through 30
another.[1] So we must examine this question by making another
beginning: if something moves itself, how and in what way does it
cause motion?

It is necessary for everything that is moved to be divisible into
parts that are at every stage divisible. This was proved previously in
our general study *On Nature*, that everything that is intrinsically 257ᵇ
moved is continuous. It is impossible, then, for the self-mover
entirely to move itself. For it would be transported as a whole, and
it would transport with the same motion, being one and indivisible
in form, and it would be altered and alter, so that it would teach
and learn at the same time, and heal and be healed with the same 5
health. Further, we have explained that the movable is what is in
motion; but it is in motion through potentiality, not through actuality, and the potential is in process to realization, and motion is the
incomplete realization of the movable. But the mover is already
actual—for instance, the hot heats, and in general what has the
form produces it in others. So the same thing will be hot and not hot 10
at the same time in the same respect, and likewise with everything
else in which the mover must have the same name as the moved.

---

[1] Reading αἴτιον ἀεὶ πρότερον with the MSS and Simplicius.

Therefore one part of the self-mover causes motion, and one part is moved.

That the self-mover does not so move that each part is moved by the other is apparent from the following considerations. In the first place, there will not be any first mover if each part moves the other.[2] (For the prior mover is more responsible for the movement than what comes next in a series, and it is more of a mover. For there were two ways of motion, by being moved by another and by being self-moved, and what is farther from the final thing moved is nearer the source of motion than what is in between.) Secondly, the mover need not be moved except by itself. Therefore it is incidental if the other part sets it in motion in return. Hence, I assume that it is a possibility that the second part does not move the first. One part, therefore, is moved, and one part is an unmoved mover. Thirdly, it is not necessary for the mover to be moved in return, but either it must move something while being unmoved or it must be moved by itself, if indeed there must always be motion. Fourthly, a thing would be moved with the motion it causes, so that the heating agent would be heated.

But surely neither one part nor more of what moves itself primarily will move itself individually. For if the whole is moved by itself, it will either be moved by one of its parts or the whole will be moved by the whole. Now, if it is moved by some part being moved by means of itself, *this* part would be the first self-mover, for if it were separated, this part would move itself, but the whole no longer would. If, on the other hand, the whole is moved by the whole, it would be incidental that the parts moved themselves. So, if their motion is not necessary, let us suppose that these parts are *not* moved by themselves. Of the whole, therefore, one part will cause motion while remaining unmoved, and one part will be moved. For only in this way will it be possible for something to be self-moved.

Further, if the whole moves itself, part of it will cause motion, part will be moved. AB, therefore, will be moved both by itself and by A alone. And since what causes motion may either be moved by another or be unmoved, and what is moved may either cause some motion or cause none, the self-mover must be com-

---

[2] Reading ἑκάτεον κινήσει ἑκάτερον.

posed of a part that is unmoved but causes motion and also of a part that is moved but does not necessarily cause motion, but may or may not cause it. Call the part that causes motion but is unmoved A, call B the part that is moved by A, and the part that moves something we shall call C; this last is moved by B but moves nothing 10 else (for even if there are several intermediaries before C, let us suppose only one). The whole ABC, then, moves itself. But if I take away C, AB will move itself, with A causing motion and B being moved, but C will not move itself, nor will it be moved at all. But 15 surely BC will not move itself without A either. For B causes motion by being moved by another, not by being moved by some part of itself. Therefore AB alone moves itself. It is necessary, therefore, for the self-mover to have a part that causes motion while it is unmoved and a part that is moved but does not necessarily cause motion, with either both parts touching each other, or one 20 part touching the other.

Now if the mover is continuous (for the moved *must* be continuous), each part will touch the other. It is clear, then, that the whole thing moves itself not by part being such as to move itself, but it moves itself as a whole, being moved and causing motion by part of it being the mover and part being the moved. For it does not move, 25 nor is it moved as a whole, but the A part causes motion, while the B part alone is moved.

A problem arises if one takes something away either from the A part—if the unmoved mover is continuous—or from the moved part B. Will the remainder of A cause motion or of B be moved? If so, AB would not be the primary thing moved by itself. For when 30 one has taken away something from AB, what is left of AB will still move itself. Or is it that nothing prevents either both parts or at least one part, the moved, from being potentially divisible but not **258ᵇ** in reality divided, so long as it is the case that if it is divided, it will no longer have the same nature? So nothing prevents the original mover from being in what is potentially divisible.

It is apparent from these considerations, then, that the primary mover is unmoved. For whether what is moved but moved by something is referred directly back to the first unmoved mover, 5 or to something moved but moving and stopping itself, in either case the primary mover for all moved things turns out to be unmoved.

CHAPTER 6

10 Since there must always be motion without intermission, there must be something everlasting which first moves things, whether one or more. And the first mover must be unmoved. The question as to whether each of the unmoved movers is everlasting has nothing to do with the present argument. But that there must be some being which is itself unmoved and devoid of all change, both
15 unqualified and incidental, but which is able to impart movement to another, will become clear from the following considerations.

Now let us grant, if you will, that some things can be at one time and not be at another time without coming to be or perishing— indeed, perhaps it is necessary, if something without parts at one time is and at another is not, that anything of that sort should at one
20 time be and at another not without *changing*. And of the principles which are unmoved but can cause motion, let it be possible that some of them are at one time and are not at another. Nevertheless, this is not possible for all. For it is clear that there is some cause of the self-movers' being at one time and not being at another. For everything that moves itself must have some magnitude, if nothing
25 without parts is in motion, but there is no need for the mover to have magnitude, from what we have said. Of the fact that some things are coming to be and some perishing and that this is happening continuously, no unmoved thing that is not everlasting could be the cause; nor could anything which causes motion to some things while other movers cause motion to other things.[3] For neither is each of them nor are all of them causes of what always and continu-
30 ously exists. For this state of affairs exists everlastingly and neces- sarily, but all those things are infinite, and they do not all exist at the same time. Hence it is clear that even if some of the unmoved
259ᵃ movers and many of the self-movers perish countless times, and others succeed them, and this thing which is unmoved moves that thing and that moves another, none the less there is something which embraces them all, and this exists in addition to each thing and is the cause of some things existing and others not, and of
5 the (sc. aforementioned) continuous change. And this causes the movement of these, while these cause the movement of the others.

---

[3] Reading τῶν ἀεὶ μὲν ταδὶ κινούντων.

Thus, if motion is everlasting, the first mover too will be everlasting, if there is one. But if there are several, there will be several
everlasting movers. But we should believe in one rather than
several, and in a limited rather than an unlimited number. For if
the consequences are the same, we must always prefer the limited   10
number. For among natural phenomena the limited and the better,
if it is possible, must exist rather than the opposite. In fact, one is
enough, which, being first of unmoved ⟨movers⟩ and everlasting,
will be the source of movement for everything else.

It is also apparent from the following consideration that the first
mover must be something that is one and everlasting. For it has
been demonstrated that there must always be motion. But if   15
motion always exists, it must be continuous, for what is everlasting
is continuous, while what is successive is not continuous. But now if
it is continuous, it is one. But a motion is one if it is the motion of
one moved cause by one mover. For if one thing, then another, is to
cause a movement, the whole movement will not be continuous,
but successive.

From these considerations one would be justified in believing   20
that there is some first unmoved mover, and also by reviewing the
beginnings of the argument. Indeed, the fact that there are some
beings which at one time are in motion and at another are at rest is
apparent. And from this fact it has become clear that neither are all
things in motion nor are all things at rest, nor are things divided
into those always at rest and those always in motion. The things   25
that alternate and have the power of both motion and rest prove
this point concerning the different possibilities. Since such facts are
clear to all, we wished to show the nature of each of the other two
types of things, that there exist some things which are always
unmoved and some which are always moved. Advancing towards
this end, and having established that everything that is moved is   30
moved by something, and that this thing is either unmoved or
moved, and if moved, then it must be moved either by itself or by
another and then another, we have proceeded to the point of
realizing that the principle of things in motion is, among things
moved, the self-mover, but among all things, the unmoved mover.

But we see that there are beings which quite evidently are such as   259^b
to move themselves: i.e. the class of animate objects and that of
animals; and these suggested the view that perhaps it is possible for
motion to arise without it ever having existed previously, because

5 we see this happening in them, for at one time they are motionless,
and then they move once more, as it appears. Now we must realize
that they move themselves with only one kind of motion, and that
they do not strictly speaking cause it. For the cause does not derive
from the animal itself, but there are other natural motions inherent
in animals, which they do not undergo through themselves, e.g.
increase, decrease, and respiration, which each animal has while it
10 is at rest and not moving with its own inherent motion. The cause
of this is the environment and many of the things which enter into
the animals, as when food causes some motions. For while it is
being digested, the animals sleep, and when it is being distributed,
they awaken and move themselves; and thus the first source of
motion is outside, which is why they are not always continuously
15 being moved by themselves. For the mover is distinct, and it is itself
moved and causes change by interacting with each self-mover. In
all these cases the first mover and cause of the thing moving itself
is moved by itself, but incidentally. For the body changes place, so
that what is in the body also changes place, moving itself through
leverage.
20   On the basis of these considerations one can be sure that if
something belongs to the class of things that are unmoved but
move themselves incidentally, it is unable to cause continuous
motion. So if it is indeed necessary for there to be continuous
motion, there must exist a first mover that is not moved even
25 incidentally—that is, if there is to be, as we have said, an unending
and undying motion in things, and what-is itself is to remain in itself
and in the same. For if the principle remains constant, the universe
must remain constant too, because it stands in a continuous relation to the principle. But it is not the same thing for something to
be moved incidentally by itself and to be moved incidentally by
another; for being moved ⟨incidentally⟩ by another is an attribute
30 belonging even to some principles in the heavens: namely, those
things that describe complex orbits, but being moved ⟨incidentally⟩
the other way is an attribute only of perishable things.
    But surely if there is something which is always such as to move
another while itself remaining unmoved and everlasting, it is neces-
260^a sary for the first thing moved by this to be everlasting also. This is
clear also from the fact that there will not be any coming to be or
perishing or change of other things unless something that is in
motion causes the change. For the unmoved mover will always

move things in the same way and with a single kind of motion, because it itself does not change at all in relation to what is moved. 5 But what is moved by something moved, which in turn is moved by something unmoved, because it occupies variable relations to things, will not be the cause of a uniform motion, but because of being in contrary places or forms it will produce contrary motions in each of the other things moved and cause them to be at one time 10 at rest and at another time in motion.

It has become apparent from our discussions what is the solution to the problem we raised at the beginning: namely, why all things are not either in motion or at rest, or why some are not always in motion while the others are always at rest, but some are at one time in motion, at another time not. The cause of this is now clear: namely, that some things are moved by an unmoved mover that is everlasting, and that is why they are always in motion; and some 15 things are moved by a moved and changeable mover, so that these things must change too. But the unmoved mover, as we have said, because it remains simple and self-identical and in the same, will move things with a single and simple motion.

CHAPTER 7

Nevertheless, the point will be more apparent if we make a new 20 beginning concerning these matters. For we must consider whether it is possible for any motion to be continuous or not, and if it is possible, what kind of motion this is and what kind of motion is primary. For it is clear that if some motion must always exist, and this is primary and continuous, the first mover causes this kind of 25 motion, which must be single and the same and continuous and primary. Since there are three kinds of motion, that of size and that of affection and that of place, which we call locomotion, it is the last that must be primary. For it is impossible for there to be increase without a prior alteration in quality; for what increases in one sense 30 increases by what is like, but in another sense by what is unlike— for contrary is said to nourish contrary. But in everything that comes to be, like is added to like. Thus this change to a contrary state must be a case of alteration. But surely if there is alteration, **260^b** there must be something causing the alteration and making (e.g.) the actually hot from the potentially hot. Thus it is clear that the

mover is not always in the same position, but sometimes nearer, sometimes farther, from what is altered. But this could not happen
5 without locomotion. If, therefore, there must always be motion, locomotion must always exist as the primary motion, and if there is a primary and a posterior kind of locomotion, the primary kind must always exist.

Further, the source of all affections is condensation and rarefaction. For heavy and light, soft and hard, hot and cold, seem to be
10 kinds of condensed or rarefied conditions. And condensation and rarefaction are aggregation and segregation, in virtue of which substances are said to come to be and to perish. And things that aggregate and segregate necessarily change place. But surely the magnitude of what increases and decreases changes also in the place it occupies.

15 Furthermore, from the following observations too it will also be apparent that locomotion is primary. For the primary in the case of motion as well as of other things is said in many ways. For that is said to be prior without which other things will not exist, while it can exist without them, and there is also priority in time and prior-
20 ity in essence. So, since there must be motion continuously, and motion would exist continuously either if motion itself were continuous or if there were a succession of motions, but much rather if it were continuous; and since it would be better for motion to be continuous than successive, and we suppose the better always to exist in nature, if it is possible, and it is possible for motion to be continuous (this will be proved later; for now let us assume it); and
25 since no other motion can be continuous but locomotion, locomotion must be primary. For there is no necessity for what moves in place either to increase or to alter, much less to come to be or perish; but none of these other motions is possible if there is no continuous motion which the first mover causes.

Furthermore, locomotion is primary in time. For this is the only
30 kind of motion possible for everlasting things. But, ⟨it may be objected,⟩ for each individual thing that comes to be, locomotion must be the last kind of motion it undergoes. For after a thing comes to be, first alteration and increase take place, while locomotion belongs only to things already completed. However, there
261^a must be another thing moving in place that is prior, which will be the cause even of generation for things generated, but which is not itself being generated, such as what begets is the cause of what is

begotten. It might indeed appear that generation is the primary kind of motion for the simple reason that the object must first come to be. Although this is the case for any individual thing that 5 comes to be, in general there must be something in motion which is prior to things that come to be, which itself does not come to be, and again there must be another thing prior to this. And since generation cannot be the primary motion—for then all things in motion would be perishable—it is clear that none of the subsequent motions is prior. By subsequent motions I mean increase, then 10 alteration and decrease and perishing. For all these are posterior to generation, so that if not even generation is prior to locomotion, neither is any of the other changes.

In general, it appears that what comes to be is incomplete and proceeds to a principle, so that what is posterior in order of generation is prior in nature. And locomotion comes last to all the things that are in process of generation. That is why some living things 15 are completely motionless due to a lack, such as plants and many genera of animals, while motion belongs to others only when they are complete. So if locomotion belongs more to the things that have more fully achieved their nature, this kind of motion would also be prior to the others in essence, both for the stated reasons and because what is moved in locomotion loses its essence less than 20 in any other kind of motion. It alone does not change its being at all, while what is altered changes its quality, and what increases and decreases changes its quantity. Quite clearly this—motion in place—is the motion that the self-mover causes above all, strictly speaking, while it is just that—the self-mover—which we say is the 25 principle of things moved and of movers and is primary among things moved.

That locomotion is the primary kind of motion, then, is apparent from the preceding considerations. What kind of locomotion is primary we must now explain. The same enquiry will also serve to justify the assumption we made both earlier and just now that some kind of motion can be continuous and everlasting. That none of 30 the other kinds of motion can be continuous is apparent from what follows. All motions and changes are from opposites to opposites: e.g. being and not being are the goals, respectively, of coming to be and perishing; contrary affections are the goals of alteration; and greatness and smallness or complete and incomplete size are the 35 goals of increase and decrease. And contrary motions are those

261^b towards contrary states. But what is not always moved with a
certain motion, if it existed previous to undergoing the motion,
must previously have been at rest. Thus it is apparent that what
changes will have been at rest in the contrary state. And the case
will be similar to changes which are not motions. For perishing and
coming to be are general opposites, and the individual case of
5 perishing is opposite to the individual case of coming to be. So, if it
is impossible for something to change in opposite ways at the same
time, change will not be continuous, but there will be a time interval between changes. For it makes no difference whether contradictory changes are contrary or not, as long as it is impossible for
the opposed states to be present in the same subject at the same
10 time: this difference is not relevant to the argument. Nor does it
matter if the subject need not rest in the contradictory state, or if
change is contrary to rest (for perhaps what is not cannot be at rest,
and perishing is a change to what is not), but only whether there
is a time interval between changes. For in that case change is not
continuous. For even in the aforementioned cases it was not the
15 contrariety of states that was relevant, but their inability to coexist
in the same subject.

There is no need to be upset by the fact that the same thing will
have more contraries than one: for instance, that motion is contrary
to rest as well as to motion toward the contrary state. We simply
need to grasp the fact that in a certain sense motion is opposed both
to the contrary motion and to rest, just as the equal and the mean
20 is opposed to both what exceeds it and what is exceeded by it, and
that it is not possible for opposites to coexist in a subject, whether
they are motions or changes. Further, in the case of coming to be
and perishing it would seem completely absurd if what had come to
be had to perish immediately without remaining for any time. So
from these cases it would be plausible to infer to other cases of
25 change, for it is according to nature that the same thing should
happen in all cases.

CHAPTER 8

That there can be some infinite motion which is one and continuous, and this is motion in a circle, we must now assert. For everything that moves in place travels either in a circle or in a straight

line or in a combination of the two, so that if one of the former is
not continuous, the combination of both cannot be continuous 30
either. That what travels with a straight and bounded motion
does not travel continuously is clear. For it doubles back, and what
doubles back on a straight path describes contrary motions. For
in respect of place, up is contrary to down, forward to backward,
and left to right, for these are the contrarieties of place. We have 35
already determined what is the single and continuous motion, that 262ᵃ
it is the motion of a single subject in a single time and in respect to
an indistinguishable form (for there were three factors: what is in
motion, e.g. a man or god, when it moves, i.e. the time, and third the
respect; and this is place or affection or form or size). Contraries
differ in form and are not one. The differentiae of place are those 5
we have mentioned. Evidence that the motion from A to B is
contrary to the motion from B to A is the fact that they halt and
stop each other if they occur at the same time. And likewise in the
case of circular motion: e.g. the motion from A to B is contrary to
the motion from A to C—for they halt each other, even if they are 10
continuous and do not double back, because contraries destroy and
prevent one another. (But sideways movement is not contrary to
upward movement.)

It is especially apparent that motion along a straight line cannot
be continuous, because what doubles back must stop, not only if it
is moving in a straight line, but even if it is traversing a circle. For 15
to travel *in* a circle is not the same as to traverse a circle. For at one
time what traverses a circle continues moving in its path; at another
time it doubles back when it arrives at the place it started from.
That this latter kind of motion must come to a halt, we have
evidence not only from the senses but also from theoretical considerations. We begin as follows. Given the three parts of a line, the
beginning, the middle, and the end, the middle stands in the place 20
of the opposite in relation to each extreme, being one in number
but two in definition. Further, the potential is different from the
actual, so that any point between the extremes of a straight line is
potentially the middle, but not actually, unless it divides the line
here when the moving body has come to a standstill and begins to
move again. Thus the middle proves to be a beginning and an end: 25
a beginning of what comes after it, and an end of what came first (I
mean e.g. if, as it travels, A stands still at B and then travels again
to C). But when it travels continuously, A is not able either to have

arrived at or to have departed from point B, but only to be there
30 at a now, and thus in no time period except that of which the now
is a division, i.e. in the whole time. But if someone claims it has
arrived and departed, A will always be at a standstill when it is
262ᵇ travelling. For it is impossible for A to have arrived at and to have
departed from B at the same time; the events, therefore, will take
place at different points of time. Therefore, there will be a time
interval between them. So A will rest at B, and similarly at all other
points, for the same account will apply to each one. Indeed, when-
5   ever the travelling body A uses the middle point B as both an end
and a beginning, it must stand still, because it is making the point
serve two functions, just as if one distinguished them in thought.
But A is the point it has departed from as a beginning, and C is the
point it has arrived at when it finishes and stands still.

This analysis can be applied to a problem, namely the following:
If line E is equal to line F, and A travels continuously from extreme
E to C, A is at point B at the same time as D is travelling from end
10 F to the other extreme G uniformly and with the same speed as A,
D will arrive at G before A arrives at C. For what set out and
15 departed earlier must arrive earlier. For A has not arrived at and
departed from B at the same time, and that is why it lags behind.
For if it arrives at and departs from the point at the same time, it
will not lag behind, but it will have to come to a standstill. We must
not, therefore, grant that while A had arrived at B, D kept moving
from end F at the same time (for if A has arrived at B, its departure
20 from that point will also have occurred, and not the same time), but
it was there in an instant rather than in a period of time.

Now in the present case it is impossible speak this way about
continuous motion. But in the case of doubling back one must use
such language. For if G travels up to the point where D was and,
doubling back, travels down again, it has used the extreme point D
25 as both an end and a beginning, one point as two. That is why it
would have to stand still. And it has not arrived at and departed
from D at the same time, for then it would be there and not there
at the same now. But surely we must not use the previous solution
to the problem. For it is not possible for the solution to be that G
is at D at an instant, and that it has not arrived at or departed from
30 that point. For it must arrive at an actual, not a potential, end. For
points in the middle are only potential resting-places, but this is an
actual one, an end as seen from below, a beginning as seen from

above, and therefore a beginning and end of both motions. There- **263ª**
fore, what doubles back along a straight line must stand still. It is
not possible, therefore, for continuous motion along a straight line
to be everlasting.

We must give the same kind of answer to those who raise Zeno's
problem: namely, that we must always cross the half-way points to 5
the goal, but there is an infinite number of these, and it is impossible to go through an infinite number; or as some put the same
problem differently, they demand that during the motion we tally
each half line before the whole as we arrive at each half-way point,
so that in traversing the whole distance it turns out that we have
counted an infinite number. And this is admittedly impossible. 10

In our earlier discussion of motion we solved the problem by
showing that time contains an infinity within itself. For there is
nothing absurd about traversing an infinite number of points in
an infinite time. And infinity exists in time as well as in length. But
although this answer suffices for the question at hand (for the 15
question is whether in a finite time it is possible to traverse an
infinite number of points or to count an infinite number), it does
not suffice to solve the true problem. For if someone sets aside the
matter of length and the question whether in a finite time one can
traverse an infinite number of points, and he enquires about time 20
itself (for time is infinitely divisible), this solution will no longer
suffice, but one must state the truth, which we have done in our
recent discussion. For if one divides a continuous line into two
halves, one treats a single point as two—for one makes it both a
beginning and an end. And both counting and bisecting produce 25
this result. And if one divides in this way, neither the line nor the
motion will be continuous. For continuous motion takes place over
a continuum, and in a continuum there is an infinite number of
halves—not actually, but potentially existing. But if one makes an
actual infinity, one will not produce a continuous motion, but will
cause it to stand still, which is plainly what happens when one 30
counts the halves; for one must count one point twice. For the end
of one half will be the beginning of the other, if one does not count **263ᵇ**
the continuous line as one, but as two halves. So one must reply to
the question of whether it is possible to traverse an infinite number
either in time or in length that in a sense it is possible, in a sense
not. For it is not possible to traverse an actual infinity, but it is 5
possible to traverse a potential infinity. For a thing moving

continuously has traversed an infinite number of points in an incidental, but not in an unqualified, sense. For the line incidentally consists of an infinite number of halves, but its essence—what it is—is different.

It is also clear that unless one takes the point of time dividing the
10 earlier and later as always in reality belonging to the later, the same thing will both be and not be at the same time, and when it has come to be, it will not be. Now the point is common to both—namely, the earlier and the later—and it is the same and one in number, but not in definition (for it is the end of the one and the beginning of the other); but in reality it always belongs to the later
15 situation. Let the time be ACB, the object D, which is white at time A, not white at time B. At time C, therefore, it is white and not white. For in any moment of A it is true to call it white, if it was white for this whole period, and in B not white, but C belongs to both times. We must not, therefore, grant that D is white at all
20 times, but rather at all times except the last now which we designated C. And this already belongs to the later period. And if not white was coming to be and white was perishing in the whole period of A, not white *has* become, and white *has* perished at C. So it was first true to say that it was white or not white at that time; otherwise, when it has come to be, the white will not exist, and when it has perished, it will exist; or it must be white and not white at the
25 same time, and in general existent and non-existent.

But if whatever exists without having existed previously must come to be, and if when it is coming to be, it does not exist, time cannot be divided into indivisible times. For if in A, D was becoming white, and it has come to be and is white at the same time in
30 another indivisible but contiguous time, namely B—if it was coming to be in A, it was not white, but it is white in B—there must be an intervening process of coming to be, so that there is also a time
264^a period in which it was coming to be. But the same account does not apply to those who reject indivisible moments, but within the very time in which it was coming to be, it has come to be and is in the last point, to which nothing is contiguous or successive. But indivisible times are successive. And it is apparent that if it was coming to be
5 in the whole time A, there is no greater time period in which it *has* come to be and *was* coming to be than the whole time in which it simply *was* coming to be.

We may rely on these and related arguments as being proper and specific to the subject-matter; and the same results would emerge from a general investigation, as follows. Everything that moves continuously, if it is not knocked off course by anything, was previ- 10 ously travelling towards the destination it arrived at by locomotion; e.g. if something arrived at B, it was travelling to B, and not just when it drew near B, but as soon as it began to move. For why should it be moving toward its goal now rather than previously? And similarly in the other cases too. Suppose that what is travelling from A, when it arrives at C, again will return to A in a continuous 15 motion. Therefore, when it is travelling from A to C, it is also travelling to A with a motion from C, so that it is pursuing contrary motions at the same time. For motions in different directions along a straight line are contrary. At the same time it is also moving from a place where it is not located. Thus, if this is impossible, it must stand still at C. The motion, therefore, is not one; for a motion 20 interrupted by a stop is not one.

Second, from the following considerations too the more general point will become apparent concerning each kind of motion. For if everything that moves, moves with one of the given kinds of motion and is at rest with the opposite kind of rest (for, as we have seen, there is no other kind besides these), what does not always move with this motion (I am referring to different forms of motion rather 25 than different parts of the whole motion) must previously have been at rest in the opposite state of rest (for rest is a privation of motion). Thus, if motions in different directions along a straight line are contrary motions and it is not possible to engage in contrary motions at the same time, what travels from A to C would not at the same time be travelling from C to A. Since it does not travel 30 both ways at the same time, but is going to pursue the latter motion, it must first rest at C. For this is the kind of rest which was opposite to the motion from C. It is clear, then, from these considerations that the motion will not be continuous.   264^b

Third, there is this argument too, more specific than the previous ones. Suppose the not white has perished at the same time as the white has come to be. Thus, if the alteration to and from white is continuous and no time intervenes, at the same time the not white has perished, both the white and the not white will have come to be. 5 For the three states will occupy the same time.

Fourth, the continuity of time does not imply the continuity of motion, but only its successiveness. For how would contrary motions such as those toward white and black share the same end-point?

The motion along the circumference of a circle, however, will be one and continuous, for nothing impossible results from it. For the
10 thing that moves away from A will be moving towards A by the same act (for it is in motion to its future destination), but it will not move in contrary or opposite directions at the same time. For not every motion to a given point is contrary or opposite to a motion from that point; but they are contrary only if they are along a
15 straight line (for this has points contrary in place—e.g. the extremes of a diameter, for they are the farthest apart), and opposite if they are along the same line. So nothing prevents such motion being continuous with no time intervening. For circular motion is from a point to the same point, but motion along a straight line is from one point to a different one. And circular motion never covers
20 the same stretches, but straight motion repeatedly does. Now it is possible for a thing which comes to be in ever different stretches to move continuously, but it is not possible for a thing which repeatedly comes to be in the same stretches. For then it would have to pursue opposite motions at the same time. So it is not possible for
25 a body to move continuously in a semicircle or in any other arc. For it would have to make the same motion repeatedly and to change to contrary motions. For the end of such a motion is not contiguous with the beginning. But in circular motion the end is contiguous with the beginning, and it alone is complete.

It is apparent from this analysis that no other motions besides circular motions are continuous either. For in all the other cases
30 things traverse the same stretches repeatedly, such as in alteration the in between and in quantitative change the intermediate magnitudes, and similarly in coming to be and perishing. It makes no difference whether we make the intermediate stages of change
265ᵃ few or many, or whether we add or take away a stage in between; in either case the motions traverse the same stretches repeatedly. Thus it clearly follows that those natural philosophers who claim that all sensible things are always in motion are mistaken. For things must move with one of the specified types of motion, and
5 especially, according to them, with alteration. For they say that things are always in flux and decay, and they call both coming to be and perishing in turn alteration. The present argument, however, has already established generally for all motions that it is not

possible to move continuously with any kind of motion except circular motion, so that something cannot move continuously with either alteration or increase. Let this conclude our arguments that 10 there is neither infinite nor continuous change apart from circular locomotion.

CHAPTER 9

That circular motion is the primary kind of locomotion is clear. For every motion in place, as we said earlier, is either circular or straight or a combination of the two. And the first two must be prior to the last, for it is composed of them. And circular is prior to 15 straight motion, for it is simple and more complete. It is not possible to traverse an infinite straight line, for there is no such infinite line, and even if there were, nothing would make such a motion; for the impossible does not occur, and it is impossible to traverse an infinite line. But motion that doubles back along a finite line is a 20 composite motion consisting of two parts, while if it does not double back, it is incomplete and perishable. For the complete is prior to the incomplete, in nature, in definition, and in time, as the imperishable is prior to the perishable.

Again, what can be everlasting is prior to what cannot be everlasting. Now circular motion can be everlasting, but no other type 25 of locomotion or change can be everlasting. For other motions must come to a standstill, and where there is a standstill, motion has perished.

It is reasonable to conclude that circular rather than straight motion is one and continuous. For there is a definite beginning and end and middle of motion along a straight line, which are all con- 30 tained in it in such a way that there is a place where the motion will begin and a place where it will end (for everything rests at its limits, whether at its beginning or end-point), but the points of a circumference are undefined. For why should any point of the curve be more of a limit than any other? Each point is at once a beginning and a middle and an end, so that the moving body is at the beginning and the end always and never. That is why in a sense a **265ᵇ** revolving sphere both moves and is at rest; for it occupies the same place. The reason is that all these attributes belong to the centre point: it is the beginning and the middle and the end of

the magnitude, so that, because it is not on the circumference, there
5 is nowhere that the travelling body can rest as having finished its
course (for it is always travelling around the middle, but not to an
end), and because this body remains in motion, the whole is, in a
sense, always at rest and always in continuous motion.

And there is a reciprocal connection: because the revolution is
the measure of motions, it must be primary (for everything is
10 measured by what is primary); and because it is primary, it is the
measure of the other kinds of motion.

Further, only circular motion can be uniform. For things moving
in a straight path travel in a non-uniform way from the starting-
point to the end-point. For the more distant anything is from its
resting point, the faster it travels. But circular motion alone has
15 neither beginning nor end naturally in itself, but they are external
to it.

That locomotion in place is the primary kind of motion is the
consensus of everyone who has discussed motion. For they attribute the principles of motion to things which cause such motion.
20 For segregation and aggregation are motions in place, and Love
and Strife cause this kind of motion. For the latter segregates, the
former aggregates things. And Anaxagoras maintains that Mind
segregates things as the first cause of motion. The same holds for
those who recognize no such cause but maintain that motion takes
place because of a void. For they too say that nature undergoes
25 motion in place (for motion resulting from a void is locomotion
and, as it were, motion in place), and of the other motions none
belongs to the first bodies but only to their compounds, according
to them. For they say that increase and decrease and alteration
result from the aggregation and segregation of indivisible bodies.
30 The same holds for all who derive coming to be and perishing from
condensation and rarefaction: they govern these processes by aggregation and segregation. Further, besides these there are those
who make the soul the cause of motion. For they say that what
moves itself is the principle of all things in motion; and the animal
266ᵃ and every animate thing moves itself with motion in place. And we
say of what is in motion in place that it is in motion in the chief
sense; but if it remains at rest in the same place, but increases or
decreases or happens to alter, it is in motion in some sense, but we
5 deny that it is in motion in an unqualified sense.

That there always was and always will be motion throughout the

32

whole of time, and what is the principle of everlasting motion, and
further what is the primary motion, and what kind of motion alone
can be everlasting, and that the first mover is unmoved—have now
been established.

<h2>CHAPTER 10</h2>

That this mover must be without parts and without magnitude let 10
us now argue, first establishing the preliminaries. One of these is
that no finite thing can cause motion for an infinite time. There are
three factors in a motion: the mover, the moved, and third, that in
which motion takes place: namely, time. These are all infinite or all
finite, or some are finite, i.e. two or one. Call the mover A, the 15
moved B, and the infinite time C. Then let D move some part of B,
which we shall call E. This will not happen, then, in a time equal to
C, for the greater motion takes more time. So the time of the
motion, call it F, is not infinite. Hence, adding to D I shall use up A,
and adding to E I shall use up B. But I shall not use up the time 20
taking away an equal portion in each case, for it is infinite. So the
whole of A will move the whole of B in a portion of C which is a
finite time period. It is not possible, therefore, for an infinite
motion to be caused by a finite mover.

Now, that a finite magnitude cannot cause motion for an infinite
time is apparent. That in general an infinite power cannot belong to
a finite magnitude will become clear from the following considera- 25
tions. Let us suppose that the greater power is the one which always
brings about an equal effect in a shorter time, as in the case of
heating or sweetening or throwing, or in general causing motion.
The thing affected, therefore, must receive some effect from the
finite magnitude which has an infinite power, and a greater effect
than from any other cause. For the infinite is greater than any finite 30
power. But now the event cannot take *any* time. For supposing
there was a time A in which an infinite force heated or pushed,
while in time AB some finite force did this, by repeatedly adding
to this force a greater finite force, I shall eventually bring about a **266^b**
completed motion in time A; for by repeatedly adding to a finite
amount, I shall exceed every definite quantity, however large, and
by taking away, I shall fall short of any definite quantity, however
small. A finite power, therefore, will cause a motion in a time equal

5 to that occupied by an infinite power. But this is impossible; therefore no finite thing can have an infinite power.

Neither indeed can a finite power exist in an infinite magnitude.
But, ⟨it may be objected,⟩ it is possible for a greater power to reside
in a lesser magnitude. True, but *a fortiori* a still greater power can
reside in a greater magnitude. Let AB be an infinite magnitude, of
which a portion BC has a certain power, which moved D in a given
10 time: namely, EF. If then I take double the amount of BC, it will
cause the motion in half the time of EF (let this be the proportion),
so that it will cause the motion in time FG. Thus, doing the same
thing each time, I shall never traverse line AB, but I shall take an
ever smaller portion of the given time. The power, therefore, will
15 be infinite. For it will exceed every finite power, given that every[4]
finite power must occupy a finite time (for if so much power takes
so much time, the greater power will move the thing in a shorter but
still determinate time, which is inversely proportional to the
power), and the sum total of the power will be infinite, just as what
20 surpasses in number and magnitude every determinate quantity.
Here is another kind of proof: we take a certain power, the same in
kind as the power in the infinite magnitude, but this one in a finite
magnitude, and it will be commensurable with the finite power in
the infinite magnitude.

25 Now, that it is not possible for an infinite power to reside in a
finite magnitude, nor for a finite power to reside in an infinite
magnitude, is clear from the foregoing considerations. But concerning cases of locomotion, we would do well to face a problem
before proceeding. If everything in motion is moved by something,
as for those that do not move themselves, how is it that some of
them will move continuously even when they are not in contact
30 with their mover, as in the case of projectiles? If the mover simultaneously moves something else, such as the air, which causes
motion by being moved, it is no less impossible for this than for
the projectile to be in motion when the first mover is not in contact
with it and causing its motion. But at the same time all things
267ᵃ will be in motion and stop whenever the first mover stops, even if,
like the magnet, it renders what it has moved able itself to cause
motion.

This is what we must reply: the first mover renders able to cause

---

[4] Reading εἴ γε πάσης of K, with Ross.

motion the air or water or whatever has a nature to move and be
moved. But the medium does not simultaneously stop moving 5
and being moved; rather, it stops being moved at the same time
its mover stops moving it, but it still causes motion. That is why
it keeps moving something else contiguous to it, and the same
account applies to its successor. But it begins to stop whenever the
power of causing motion becomes continually less in each contiguous thing. And finally it stops when the previous body no longer
makes the next one a mover, but only something moved. And these 10
things must stop simultaneously, the mover and the moved, as well
as the whole motion. Now this motion comes to be in things that
can be at one time in motion and at another time at rest, and it
is not continuous, though it appears to be. For it occurs either in
successive things or in things in contact; for there is not a single
mover, but a series of contiguous ones. That is why there takes 15
place in air or water the kind of motion some call 'recirculation'.
But it is impossible to resolve the problem in any other way than
the one mentioned. In recirculation everything would simultaneously cause motion and be in motion, and thus also stop simultaneously; but as it is, there appears to be one thing in continuous
motion. What keeps it in motion then? Surely not the same thing 20
that set it in motion.

Since in the realm of beings there must be continuous motion,
and this is a single motion, it must be a single motion of a certain
magnitude (for what does not have magnitude does not move) and
of a single thing moved by a single mover (otherwise it will not be
continuous, but a series of motions contiguous to each other and
separate); as to the mover, if it is single, it will cause motion either
by being moved or by being unmoved. Now if it is moved, it will 25
itself have to follow suit and to change, and by the same token to be
moved by something, so that the series will stop when we arrive at **267**^b
a motion caused by something unmoved. For this unmoved thing
will not have to change along with the others, but it will be able to
cause motion always (for to cause motion in this way does not
involve work), and this kind of motion alone or especially is
uniform. For the mover does not experience change at all. And 5
what is moved must not experience change in relation to the mover
either, if its motion is to remain constant. The mover, then, must be
either at the centre or on the circumference of a circle, for these are
the principles of a circle. But things nearest the mover move most

35

swiftly. And such is the motion of the circumference. That, therefore, is where the mover is.

There is a problem as to whether it is possible for something in
10 motion to cause continuous motion in a way different from what
pushes repeatedly, by the series of impulses being continuous. For
such a mover must either always push or pull or both, or another
thing receiving an impulse from a series of intermediaries must do
so, as was mentioned previously in the case of projectiles, on the
grounds that since the air is divisible, a different portion always
15 causes motion by being moved. In neither case can there be a single
motion; there can be only a contiguous motion. The only continuous motion, therefore, is that which the unmoved mover causes.
For, by always remaining in the same state, it will also hold the
same relation continuously to the moved.

Now that these points are established, it is apparent that it is
impossible for the first and unmoved mover to have any magnitude.
For if it has magnitude, the magnitude must be either finite or
20 infinite. That there cannot be an infinite magnitude has already
been proved in the *Physics*. That a finite magnitude cannot have
infinite power, and that something cannot be moved for an infinite
time by a finite magnitude, has just now been proved. But the first
25 mover causes everlasting motion for an infinite time. Plainly, then,
it is indivisible and without parts, and it has no magnitude.

COMMENTARY

CHAPTER 1

250^{b}11–15

Aristotle is concerned with the question of the beginning and end of motion, because it has immediate implications for cosmology. If motion comes to be or ceases to be, the natural world must correspondingly come to be or perish. For Aristotle, nature entails motion of a certain sort: namely, motion that originates with the natural body itself (see Introduction). If that motion were to cease, there would then be no natural world. Motion, then, is coextensive with nature and with the existence of the cosmos (for the meaning of this term, see Introduction), which is a system of natural bodies, and also with time itself (*Phys.* IV. 11). If we can establish the limits or limitlessness of motion, we can determine whether our cosmos is transient or everlasting. Aristotle's main objective, however, is not to determine the duration of the cosmos but the cause of its duration, which is to say: the cause of continuous natural motion.

We should pause here a moment to consider the merits of Aristotle's initial approach. Although the present discussion seems perfectly reasonable in the rather ethereal domain of speculative cosmology, it is striking that before Aristotle no one had addressed the problem of cosmic motion in a comprehensive way. As he notes, many assumptions were made about motion. But among the Presocratics we find not a trace of a discussion of principles in general. And even Plato in the *Timaeus* fails to discuss the theoretical options. He simply assumes that some sort of change has always taken place in the world of becoming (if we take his discussion of the origin of the cosmos at face value, rather than as an allegory: *Tim.* 48b, 52d–53b; Vlastos 1939, 1965*a*; see below on 251^{b}17–18). Perhaps the nearest predecessor to the present discussion is found in *Laws* X, where Plato discusses general questions of motion, rest, and causation. But Plato's focus is more on theodicy than on cosmology, so his discussion lacks the metaphysical and scientific rigour of Aristotle's treatise. Aristotle, by contrast, wishes to consider all theoretical possibilities and to identify those who hold each one and what reasons support or undermine the view. Whatever we may think of the subsequent discussion, we must acknowledge the fact that Aristotle has raised the discussion of cosmology to a new level of generality and rigour just by this theoretical prolegomena. Compare his similar but brief discussion of the problem in *Cael.* I. 10.

250^{b}15–17

'that there is motion is the view of all who discuss nature': so far as we can see from the fragments, the early philosophers of nature (i.e. the Presocratics excluding the Eleatics, who denied the existence of change) did not explicitly take a stand on the existence of motion. Nor is such an abstract topic characteristic of Presocratic speculation. Aristotle's characterization is based on an inference to what their cosmogonies presuppose (cf. Burnet 1892/1930: 12).

250^{b}16–17: 'their whole study concerns coming to be and perishing': Aristotle runs the risk of misleading us by saying that the natural philosophers all focus on coming to be and perishing. The pluralists (see Introduction) explicitly rejected coming to be and perishing as an impossibility (e.g. Empedocles B8–12; Anaxagoras B17). It is true that they dealt with the *phenomena* of coming to be and perishing, and that in some sense all held that the cosmos comes to be (and some of them held that it would perish). According to Aristotle, some of the phenomena of coming to be and perishing do indeed constitute real instances of coming to be and perishing. But the pluralist philosophers would not agree that at the level of philosophical (or physical) analysis there is any real coming to be or perishing.

250^{b}18

'unlimited worlds': the view that there are multiple worlds was clearly held by the atomists, Leucippus and Democritus (DK 67A1, 68A40). Doxographical reports also attribute the view to Anaximander (DK 12A14, 17), although there is debate about whether the report is right, and, if so, whether the worlds were

38

contemporaneous or successive (see Kahn 1960: 33–5, 46–53; KRS 122f.).

'a single world, or one that does not always exist': Ross and other commentators wish to emend the text here because the contrast between a single world and one that is not everlasting does not have any immediate implications for the eternity of motion. Yet Aristotle elsewhere uses what appear to be at first sight illogical and irrelevant contrasts. For instance, he begins *Met.* Λ by developing a classification of substances into sensible on the one hand and immovable on the other. Closer scrutiny reveals that there is a deeper principle of division in that case, one that is relevant to his enquiry. Here also we can glimpse a possible basis for contrast. Some philosophers have a single world, others have a repeating world, where there is only one world at a time, but one which ceases to be and then comes to be again. The contrast Aristotle seems to have in mind is between a single *continuous* world and a (single) non-continous world. The contrast is poorly stated, but could not be rescued easily by an emendation. It might seem misleading to treat the reiterations as manifestations of a single world (though Aristotle does effectively analyse the situation as consisting of changing states of the same world, *Cael.* I. 10, 280ᵃ11–23), but it is definitely not an instance of plural coexisting worlds of the kind the atomists envisaged. An example of the repeating world is that of Empedocles, whose cosmos comes to be repeatedly. An example of a single, non-repeating world is that of Anaxagoras. Significantly, these are the two cases which Aristotle discusses in the next paragraph as paradigm examples. And both imply principles of motion.

Ross adds *ē aei* and translates, 'those who think that there is but one universe [sic; in fact, the topic is really worlds or world orders (*kosmoi*) within the universe (*to pan*)] and make it eternal or not eternal, make movement eternal or not eternal accordingly'. But what Presocratic cosmologist would satisfy his alternative of an eternal world? Heraclitus presumably could, if we follow the majority of present interpreters of his B30; but Aristotle does not read Heraclitus this way (*Cael.* I. 10, 279ᵇ12–17, with Simpl. *in Cael.* 94.

4; cf. *Cael.* II. 1, 298^{b}24–33); indeed, at the beginning of the former
De Caelo passage he says explicitly, '*All* [the natural philosophers]
say the world came to be, but once it has come to be, some make it
everlasting, some make it perishable . . .'. After a process of coming
to be, the world can be everlasting (on the view in question), but it
is not eternal. Thus Ross's interpretation seems to be unlikely.
(Melissus's world is everlasting (B1, B2), but it is not a *natural*
world in which change exists.)

250^{b}23–251^{a}5

'Now if it is possible for there to be a time at which nothing is
moving': sc. and another time at which something is moving. One
could hold that motion is everlasting, as the advocates of plural
worlds do. But one could also hold that motion comes to be at some
given time in the past; or one could hold that there are times of
motion alternating with times when there is no motion in the world.
Aristotle presents these latter two possibilities in the present paragraph. Anaxagoras holds the first view, Empedocles the second.
Aristotle refers to Anaxagoras B1 and B13 and quotes Empedocles
B17. 9–13 = B26. 8–12. Aristotle's interpretation of Empedocles as
having two times of rest (rather than just one, when the elements
are perfectly unified in the Sphere) is controversial, for some recent
commentators argue that there is no break between the emergence
of the elements from the Sphere and the process of reunification of
the elements into the Sphere (Tannery 1887/1930, von Arnim 1902,
Hölscher 1965, Solmsen 1965, Long 1974, KRS, Osborne 1987; for
the traditional view O'Brien 1969, 1995, Barnes 1979*b*, Wright
1981, Graham 1988, Inwood 1992). In any case, Aristotle assumes
a distinct reign of Love and reign of Strife (*GC* II. 6, 334^{a}5–7), with
periods of rest between. The traditional account of Empedocles'
cosmic cycle is as follows: (1) period of increasing Love, forming
compounds; (2) the Sphere (*Sphairos*), in which Love rules completely and all elements are joined in a perfectly homogeneous
whole; (3) period of increasing Strife, which separates the elements
from the Sphere; and (4) period of complete separation, which I
take to mean the complete stratification of the four elements into
concentric cosmic shells: earth, water, air, fire. Rest will occur at
stages (2) and (4). For the most part the controversy is not crucial

to Aristotle's argument. What is important is that times of cosmic motion alternate with times without motion, and clearly Empedocles represents this view, whether there are one or two periods of rest in the cosmic cycle. But see on 252^{a}7 below.

251^{a}5–8

Here Aristotle gives the first glimpse of why the question of cosmic motion is important. By answering the question as to whether cosmic motion is everlasting or has a beginning in time or is intermittent, we can come to a better understanding of nature, and in particular, of the ultimate source of motion. We shall be able to infer from the presence of motion in the cosmos to the character of the first principle of motion.

251^{a}8–17

'the definitions we have already laid down': Aristotle refers back to his definition of motion in *Phys.* III. 1, 201^{a}10–11. As Ross notes, Aristotle seems to regard his present treatise as separate from the *Physics* proper, consisting of Books I–IV (Simplicius here refers to Books I–V as the *Physics* proper). In Book III Aristotle treats motion as having several species, with corresponding definitions. Not only is motion the actuality of the potentially movable as such, but alteration (or change of quality) is the actuality of the alterable as such, increase and decrease are the actuality of what can increase and decrease as such, and coming to be and perishing are the actuality of what can come to be and perish as such (201^{a}11–15). Here Aristotle refers to specific cases of motion with their corresponding subjects. He also distinguishes, in passing, the agent—i.e. that which has the power to cause change, e.g. to burn something— from the patient—i.e. that which has the power to be changed, e.g. to be burnt.

251^{a}15–17: 'So there must be something burnable . . .': Aristotle infers from the definition of motion and the examples that motion presupposes a subject which has the potential to move things and an object that has the potential to be moved. Hence motion presupposes movers, active and passive. Thus the movers, like the motions, must either come to be at some time or be everlasting. But this argument will be compelling only if we have reason to believe that there is motion in the world. This point Aristotle has not really established. Earlier ($250^b18–23$) he argued that the natural philosophers assume that there is motion; but that point is valid only *ad hominem*, and only for one subset of philosophers (it does not apply to the Eleatics). Aristotle could appeal to *Phys.* I for some arguments to the effect that there is motion. But since at present he is offering a new argument, he should defend the existence of motion itself before explicating the implications. Ultimately, Aristotle holds that there is no need to give a theoretical argument for motion: it is just a fact of experience ($3, 253^a32–^b2$)—a very reasonable position. But if he is relying on that assumption, he should tell us so.

One further problem suggests itself: why must we identify an active and a passive mover in each situation? Could there not be a self-contained self-mover? Plato held that the original motion was self-motion (*Phdr.* 245c–e, *Laws* 894cff.). Aristotle will deal with this question in Ch. 5, but for now he finesses it.

251ᵃ17–28

This argument seems to go as follows:

(1) Suppose there is a first change C.

(2) Thus there is a first mover and a first moved (we posit one of each, M_1 and M_2, respectively).

(3) M_1 and M_2 either come to be or they are everlasting (from $^a16–17$)

(4) Suppose M_1 and M_2 come to be.

(5) Then there is a prior change: namely, the coming to be of M_1 and M_2.

(6) Thus, if M_1 and M_2 come to be, there is a change prior to M_1's acting on M_2.

(7) Let M_1 and M_2 be everlasting.

(8) Let M_1 act on M_2 at time t_1.

(9) At time t_0 prior to t_1, M_1 was at rest.

(10) There must be some cause of M_1's being at rest.

[(11) There must be some change to cancel the cause of M_1's being at rest.]

(12) Thus, if M_1 and M_2 are everlasting, there is a change prior to M_1's acting on M_2.

(13) Thus there is some change prior to M_1's acting on M_2 (3–12).

(14) But C just is M_1's acting on M_2.

(15) Thus C is not the first change,

which contradicts the original hypothesis. The argument is a dilemma in which we see that on any supposition, there must be a change prior to the alleged first change. The most puzzling step of the argument is (10). Aristotle evidently finds it inexplicable why M_1 and M_2, which are of a nature to interact (he will elucidate this relationship in his next paragraph), do not interact. There must be some cause—some obstacle, we might say, to their interacting. In general, however, Aristotle does not require any cause for rest, though he does require a cause for motion or change. Focusing on the claim that rest is a privation of motion, ^{a}26–7, Aquinas notes that a privation exists as a (presumably acquired) condition of a subject only through a cause. But, on the contrary, rest is not a privation like blindness, which can be taken either as an acquired condition or as a loss of a capacity: cf. *Cat.* 10, 12^{a}26 ff. For what condition has been acquired, what capacity lost? Simplicius 1128. 18–25 invokes privation in the sense of a capacity, but this makes it difficult to see in what sense rest needs a cause. There is a still more fundamental problem: even granted (10), it is not clear that the cause in question constitutes a *change* (Aristotle's four causes are notoriously broader in scope than our modern scientific notion of cause: see on 252^{b}4). Now it would be more in keeping with Aristotle's own principles to require that some cause operate as an obstacle to M_1's action, or as an explanation of M_1's inactivity as a cause, than that it act as a cause of rest. Then we would need to emphasize not (10) but (11), the need for a cause to eliminate the obstacle or to 'turn on', as it were, M_1 (compare his strategy below at 255^{b}13 ff.). But this move immediately supplies the change that is needed to complete the argument: namely, the change of removing the obstacle. Accordingly, I insert (11) as a sequel to (10). This interpretation may be what Aristotle has in mind in this compressed account; but even if not, it provides a viable path to his conclusion. At this point the argument becomes very much like a classic argument introduced by Parmenides, B8. 9–10: 'What

43

need would have stirred it, later or earlier, starting from nothing, to grow?' That is, given that M_1 and M_2 have lain dormant for an infinite amount of time (since they are everlasting), why should they suddenly leap into activity? (See below on 252ᵃ14.)

Aristotle seems to have another reason for rejecting a primeval state of inaction, one which he does not articulate here, though it could inform the present argument. Given that time is everlasting, there will be an infinite period before the first instant of change when cosmic motion began, if it did. But according to Aristotle's 'Principle of Plenitude', in an infinite period of time all possibilities are actualized (*Cael.* I. 12, 281ᵇ20–3, 25–7; Hintikka 1973: ch. 5; the term is from Lovejoy 1936; Judson 1983: 225–8, however, argues that the principle applies only to natural capacities of the elements). Hence, if M_1 and M_2 had been in existence and had had their present capacities, they should have interacted before any given time *t* at which they did, by hypothesis, first act. If, among the causes of M_1's and M_2's being at rest ((10) above) we count the lack of a capacity—manifested by their failure to interact over an infinite period of time—we have an argument similar to the one he expresses here.

251ᵃ28–ᵇ10

But the argument in the previous paragraph needs explication. Why would the movers and movables have to interact? We must distinguish between things that always act the same way (those with 'irrational' potencies, such as fire, which always burns and never cools) and things that can act in opposite ways (those with 'rational' potencies, such as a doctor, who can either heat or cool a patient to achieve his goals: *Met.* Θ2, 5). The former must act automatically when the active potency meets the passive (5, 1048ᵃ5–7); e.g. if fire comes into contact with wood, it ignites the wood. (For a valuable discussion of the assumptions see Cornford 1937: 162–77 on Plato's *Timaeus*.) The movers and movables that on this hypothesis constituted the cosmos before motion occurred would have only irrational powers. Thus, if the world was so arranged that it consisted of movers and movables in contact with each other, the movers would have moved the movables automatically. Hence, we must suppose that either something changed from being a non-mover to being a mover, or something changed from being a non-movable to being a movable, or some latent power of the mover or the movable was 'turned on', or the mover and the movable were brought together by some change of place, or some obstacle to their interaction was removed. In any case, we must assume some motion prior to the first movement of the world.

Should Aristotle confine his investigation to cases of irrational powers? One of the theories he is opposing—namely, that of Anaxagoras—starts cosmic motion with a rational agent (Nous—Mind or Reason itself) giving the initial push. Here, it seems, Aristotle could still urge his objection by pointing out that Anaxagoras needs to tell us why Nous initiated motion at this time rather than at some other. 'Because it decided to,' Anaxagoras might answer evasively. But why, in the first place, did it decide to do it, and why, in the second place, did it decide to do it now? If its action was rational, Nous would need some reason for acting, and further, some reason for acting now rather than later; something, then, must have changed, either in the universe as a whole or in the cosmic Mind. If, on the other hand, there is no reason for Nous to initiate motion now, perhaps Nous acted on an irrational impulse. Now it does not immediately follow that Nous is an irrational power ('irrational' here is equivocal between 'acting without reason' and 'lacking the capacity to reason'); but it does appear that some external cause would have to activate Nous, so that his objection to a first motion might still apply to the act of a rational agent. A critic might reply that to say Nous acts on impulse is not to supply a cause for the action of Nous at all. True, but it invites us to look for a cause—at least for the change in psychological state—and this brings us back to Aristotle's objection.

251ᵇ10–28

This paragraph is a digression or a later insertion, as we see from the beginning of the next paragraph. Here Aristotle argues that time is everlasting, and since time is a property of motion, it must follow that motion is everlasting. He presupposes his theory of time in *Phys.* IV. 10–14. His theory is less than perspicuous at some points; see Hussey (1983: 150 ff.) for some detailed discussion of the theory and its problems. In any case, to his previous argument that every change presupposes some prior change, Aristotle adds this argument, that the existence of time itself presupposes everlasting change (best exemplified by the everlasting motion of the heavenly bodies).

According to Aristotle, time is 'number of motion in respect of the before and after' (IV. 11, 219ᵇ2). I take this to mean that time is the quantity which motion exhibits in the dimension marked out by predications of 'before' and 'after'. The definition presupposes the existence of motion, and Aristotle repeatedly marks the close relation between time and motion. Moreover, the relation seems to be one of ontological dependence: 'That time is neither motion nor independent of [literally: 'without'] motion, is clear' (IV. 11, 219ᵃ1– 2). If, then, time is everlasting, as Aristotle maintains, he must be committed to having motion be everlasting. Of course, in the order of causation, the eternity of motion explains the eternity of change, and not vice versa.

There is a danger, however, in leaning too heavily on Book IV in explicating the present argument. For at IV. 13, 222ᵃ29–30 Aristotle asks, 'Will time then fail?', and answers only with another question, 'Or will it not, since there is always motion?' If there is not a strong independent argument for the eternity of time, the present argument will be circular. The argument at ᵇ14 is merely dialectical and cannot supply the need. The main argument must come at ᵇ19. On Aristotle's rhetorical question, 'how will there be a before and an after if there is no time?' see also remarks on ᵇ19.

251ᵇ14: 'all ... clearly agree': again (cf. on 250ᵇ15), Aristotle must be dealing with presuppositions rather than explicit statements.

251ᵇ17–18: 'But Plato alone generates time': *Tim.* 38b–c. It has been controversial since Aristotle's time whether Plato's remarks about the creation of the cosmos are to be taken literally or symbolically. Xenocrates read the passage allegorically and won over many Platonists to his view (Aristotle, *Cael.* I. 10, 280ᵃ30–ᵇ2, with Simpl. *in Cael.* 303. 34–5; Plutarch, *De animae procreatione* 1013A– C; Philo, *De aeternitate mundi* 13–16, 25–7, 38). But in the passage just cited, Aristotle took Plato literally, and was followed by Epicureans and some Platonists (Cicero, *Nat. D.* 1. 19ff.; Plutarch, *De animae procreatione* 1013A–C; Proclus, *In T.* I. 276 Diehl). Theophrastus did not know whom to believe (frr. 28, 29 Wimmer).

In modern times the allegorical interpretation was universal until about fifty years ago (Zeller 1876/1881; Archer-Hind 1888: 37–41; Theiler 1925/1965; Taylor 1928: 66–9; Cornford 1937: 37–9; Cherniss 1944: 420–30). Now the literal reading has emerged as dominant (Vlastos 1939, 1965a; Hackforth 1959; Mohr 1985: 40–1, 178–83; Robinson 1987; Zeyl 1987). The latter view seems to be correct: note that Plato gives a formal argument for creation (*Tim.* 27d–28c) before he makes disclaimers about the limitations of scientific speculation (29c–d), and that the disclaimers apply properly only to the specific reconstruction of the cosmos, not to the fact of creation.

251ᵇ19–23: For Aristotle's account of the now, see *Phys.* IV. 11, esp. 220ᵃ5, and ch. 13. The now has a kind of curious status as both a subjective reference point and an objective point on the time continuum (though it is not a *part* of time, since only extended quantities can be parts of extended quantities: IV. 11, 220ᵃ18–20). The now gets its name apparently from the fact that at some time an instant of time is, was, or will be 'now' for us; but this suggests that the now is somehow dependent on our consciousness for its existence. Aristotle seems to fall into the trap of saying that for its full existence time depends on soul (IV. 14, 223ᵃ21–9, with Ross's criticisms, 68). On the difference between Aristotelian and modern semantics with regard to 'now', see Hintikka (1973: 85–6).

The now is a mean (*mesotēs*) not merely in the sense of being a mid-point, but in the sense of being like a geometric mean. In the geometric mean *a* is to *b* as *b* is to *c*. Here the past time is a beginning relative to the present, while the present is a beginning relative to the future.

Aristotle's argument seems to be the following:

(1) The existence of the now is a necessary condition for the existence of time (from Bk. IV).
(2) The now is intermediate between past and future.

Point (2) is, I take it, justified at this point merely by an appeal to present experience.

(3) Thus, it is the end of the past and the beginning of the future.
(4) Thus, there is time before now (namely, the past) and after now (the future).

At this point we must suppose that Aristotle is not just taking the moment that happens to be present; he must maintain that the same holds true for any arbitrary moment of time. We make an inductive leap:

(5) For any arbitrary moment of time, whether past or future, (4) holds true.
(6) Thus, there is no beginning or end of time.
(7) But there is time if and only if there is motion.
(8) Thus, motion is everlasting.

Premiss (1) is an expression of the fact that time is a continuum the dimensionless cross-sections of which are nows. The now is a limit of time (IV. 14, 222^{a}12). Time and the now mutually entail one another (219^{b}33–220^{a}1). The theory of time in Book IV also entails premiss (2) (see 222^{a}10–13). And premiss (3) is little more than an explication of what (2) means. If, then, we grant Aristotle his theory of time, we must accept the first three premisses of the argument.

(4) follows from the first three premisses. But trouble arises in moving to (5). Aristotle does indeed speak freely as if it is simply a property of the now that it always marks a boundary between the past and the future. But if the question of whether there were an infinite past and future were in question, it seems that the objector to infinite time should be free to challenge the assumption that the now is a boundary in both directions. We might put the objection in this way. From a subjective point of view, it is true that my nows are always boundaries between the past and the future I perceive. But this property of being a boundary is not part of the essence of a now; it is only a contingent fact of my subjective experience. One could give a perfectly reasonable Aristotelian account of time without any subjective component. The now is simply a datable point on a time continuum. It is not part of its essence that the now joins past and future; by nature, it is merely a cross-section of a time continuum. If that continuum has a first moment, then there will be no past before the first now. It will provide a boundary for the future, but none for the past. Aristotle's phenomenological discussion of the now seems designed to block such a possibility; but the objector should refuse to accept that discussion. He might well maintain that such an account is un-Aristotelian in its subjective point of view. The existence of time does not depend upon the

existence of conscious subjects, any more than does the existence of space.

Aristotle's question, 'how will there be a before and an after if there is no time?' at ^{b}10–11 could be an anticipation of the present argument. It could also point to a more subtle argument: if, for instance, there were some time in the distant past when time did not exist, say t_0, what sense would it make to say, 'What happened before t_0?' For the question would be nonsensical. But in fact, it always makes sense to ask, 'What happened before (or after) t?' for any t. Thus there can be no time before which or after which there is no time. Compare the condensed argument at *Met. Λ*6, 1071^{b}7–9. Such an argument presumes that what is conceivable or perhaps linguistically well formed and meaningful reveals the way things really are—a presumption evident in the *Categories* and elsewhere throughout the Aristotelian corpus.

There is, however, one serious metaphysical challenge in Aristotle's question, 'how will there be time if there is no motion?' While we can conceive of a motionless world, e.g. that of Melissus, can we make sense of the notion that there is time at all in that world? In point of fact, we do measure time in terms of natural cycles, e.g. years, months, and days, corresponding to changes in the position of the earth relative to the sun and the moon relative to the earth and sun. If there were no such cycles, how could we measure time? And indeed, what would it mean to say there was time? We might imagine ourselves observing Melissus's world. But then we seem to be putting his world in a larger universe in which there is change (namely, our own) (cf. Berkeley, *Principles* 23, on the problem of imagining without an observer). We seem to need some frame of reference to judge that a world is timeless; such a frame of reference is possible only in a world (even a subjective, inner one) in which there is change. But, by hypothesis, Melissus's world is the whole universe; i.e. it has no place for an observer with a temporal frame of reference. But the fact that there is no room for an observer, far from making Melissus's position invulnerable, seems to render his theory unintelligible: even if it were true, no one could ever know it. Thus Aristotle is right to point out the problem of a universe without change. We can well conceive of a universe without *apparent* change; but we seem to need to posit some change to account for the very possibility of marking the passage of time relative to which no change is observed.

By analogy to the argument in 251^{a}17 ff., intermittent motion would presuppose not only a motion before the first motion, but also a motion after the last motion. In the present case the thing that causes the perishing must itself perish after all other motion has ceased. But this seems to require that some other agent cause this first agent to perish, etc.

Aristotle's argument here is compressed and somewhat difficult to reconcile with the argument about the coming to be of motion (as Cornford rightly points out; Ross, on the other hand, thinks it is easy to fill in the gaps). The earlier argument was a dilemma in which the two disjuncts of (i) the mover and moved coming to be or (ii) their being everlasting are examined separately. Here the corresponding case (ii), in which the mover and the moved continue to exist, is not considered, unless it is in the sentence, 'For being moved and being able to be moved do not stop at the same time . . .' (^{b}31–2), as Cornford notes. In that case we must suppose that the mover and the moved cease to have the capacity to move (or perhaps to interact); but that loss of capacity requires a cause, etc.

One further problem is the fact that ceasing to exist does not seem to be perfectly symmetrical with coming to exist. While we might demand that the original mover's coming to exist needs a cause, we might not be inclined to demand the same of the last mover's ceasing to exist. In modern thermodynamics the law of entropy requires an explanation for increasing order, but not for increasing disorder, since that is the natural course of events. The point is not dependent on modern physics, either: a similar problem is raised by Plato to the immortality of the soul (*Phd.* 87b–e). Even if by transmigration a soul inhabits many bodies, it may grow old in time like a tailor who has worn out many of his own suits but who in the end himself wears out. Cebes draws on a common-sense comparison between ceasing to exist and human death. The existence of a human being requires an external cause (the parents), but his death requires no such cause, only time. In light of this potential asymmetry, Aristotle should perhaps have devoted more effort, not less, to his argument on the impossibility that motion should perish.

252ᵃ4: 'motion at one time, rest at another' (*hote men ēn hote d'ou*): Does Aristotle mean the phrase to stand for (a) the model in which motion begins only once (as in Anaxagoras) or (b) the model of the cosmos in which motion and change alternate (as in Empedocles), or to cover both? Or does he mean to indicate (c) a case in which motion always existed but finally ceases: i.e. the inverse of Anaxagoras's cosmos (and is that why the positive limb of the antithesis comes before the negative)? Furthermore, since the present paragraph could be taken to prove (c) and the previous argument to have proved (a), perhaps (d) Aristotle thinks that by ruling out (a) and (c), we have *ipso facto* ruled out (b), which involves repeated cases of starting and stopping.

Aristotle has used the phrase *hote men, hote de* in 251ᵃ24–5, where the context did not involve (b), but only, apparently, (a). Thus, it does not appear that Aristotle is necessarily addressing (b) at all. On the other hand, he does not in what follows seem to make a clear transition to addressing (b) separately. So it may be that he accepts something like (d), without, however, having singled out alternating motion and rest as a special case. That is, if his argument has the perfectly general consequence that there cannot be a time when there is no motion, he need not challenge (b) at all. The interpretation of the passage depends crucially on how we take the whole argument of the chapter; see further discussion in Concluding Remarks to this chapter. For now, I take it that the intermittent motion Aristotle is discussing does not imply alternating motion and rest, but rather the possibility that at some time there is motion, at some time rest, without specifying how many times motion and rest might recur, or in what order they might occur.

252ᵃ5: 'sheer fantasy': Aristotle uses *plasma* in similar expressions in *Cael*. II. 289ᵃ6 and ᵇ25 as a strong assertion of absurdity (cf. also *plasmōdes* ('fantastic'): 5, 257ᵃ23).

The beginning of the new paragraph constitutes a crucial turn for understanding Aristotle's overall argument. See Concluding Remarks to this chapter.

252ᵃ7–10

'Love and Strife rule and cause motion in turn, and rest in the intermediate time': it is not very clear what Aristotle is saying in this compressed account. Aristotle makes a similar remark at 250ᵇ26–9 (see on 250ᵇ23 above). But whereas in the former passage he stressed the motion and rest of the cosmos itself, here he discusses the activity of Love and Strife. (i) Does e.g. Love rule and cause motion during part of the cycle, then rest when Strife is in the ascendant? Or (ii) does Love rule part of the time, cause motion for a longer stretch of the cycle (i.e. when it is not necessarily ruling, but is still active), and rest when there is no motion in the cycle, presumably when the elements are perfectly united in the Sphere and perfectly separated in the chaos brought on by Strife? Whatever Aristotle's view, both interpretations have problems. As to (i), it appears that ruling and causing motion are not coextensive: B35 portrays a gradually expanding sphere of influence for Love (on the interpretation of the fragment, see Graham 1988: 308 and n. 39). It is possible, however, that e.g. when Love's dominion is expanding, Strife is not causing any *new* separations to occur, and in this sense has ceased to cause motion, even though its law still prevails in part of the world. As to (ii), ruling and rest are not mutually exclusive: the Sphere seems to be at perfect rest, with no strife in its limbs (B27a), 'rejoicing in circular solitude' (B27. 4, B28. 2). Although the acme of Strife's rule is less well documented, presumably Strife might rest while the elements are completely separated. There are times of activity, dominance, and rest of the two forces Love and Strife, but their relation requires a more subtle handling than Aristotle gives them, and they cannot simply be identified with periods of motion and rest in the cosmos. On the alternation of rest and motion in Empedocles see now O'Brien (1995).

In any case, Aristotle's objection seems to be that one who says it is natural for motion and rest to alternate owes us a further account of how this takes place. If motion succeeds rest in the cosmos, there must be some reason for the origination of motion. Now, in fact, Empedocles does supply a reason, in terms of the fluctuating dominion of Love and Strife. At this point, however, Aristotle will presumably object that the account given is not appropriate for physical explanation: it is a psychological or social account of the interaction of two personifications, not an account of natural motions. Or, if we look at Love and Strife as forces rather than as personas, we have merely pushed back the explanation by noting that they act intermittently. Why do they do so?

252ª8: 'in turn' (*en merei*): for the meaning, see LSJ s.v. μέρος II. 2, cf. ª20 and ª28 (and earlier 250ᵇ27) echoing Empedocles B17. 29, 'they [the four elements plus the forces of Love and Strife] rule in turn as time rolls round'; i.e. they take turns ruling. The phrase is picked up also by Plato, *Sph.* 242e5, in discussing the 'Sicilian Muses'; for similar phrases in the political context of alternating rule, see Bonitz 455ᵇ13–23. The principle of alternating rule was a centre-piece of democratic political reform, and was taken over as a defining concept of health in Greek medical theory (Alcmaeon B4) and of cosmic justice in philosophy. See Vlastos (1946: 80 n. 105; 1947: 158 *et passim*).

252ª10: 'a single principle of motion' (*mia archē*): Aristotle probably intends a contrast with Empedocles' two principles of motion, as commentators universally take the passage. However, *archē* can mean 'beginning' as well as 'principle', and it is possible that he has in mind 'a single beginning of motion', in contrast to Empedocles' recurring cycles of motion.

252ª11–12

'there is nothing disorderly in things which happen by or according to nature': this point echoes a well-established Aristotelian principle that nature is a cause that acts for an end (*Phys.* II. 8, 199ᵇ32–3), that nature does nothing in vain (e.g. *Cael.* I. 4, 271ª33), that natural events occur always or for the most part (*Phys.* II. 8, 198ᵇ34–6). Implicit in all these characterizations is the more general claim that nature acts in a regular, orderly way, an assumption that guided all early Greek natural philosophy. Unlike the Presocratics, however, Aristotle sees the regularities as having a place within a teleological framework.

252ª14–16

There seem to be two problems with supposing that motion occurs after an indefinite time when there was no motion: (1) the change presupposes an undefinable relation between infinite periods of rest and motion; (2) there would be no explaining why this change happened now rather than at some previous time. As to (1), if time has no beginning, then an infinite time will pass before any datable beginning of motion; and if motion continues ever after (as it does for Anaxagoras, B12, B13), since time has no end, motion will exist for an infinite time. But according to the previous sentence (ᵃ13–14), there can be no ratio between infinite quantities. Note that even if the time of motion were taken as finite, Aristotle would have objected on the similar ground that there could be no ratio between an infinite and a finite quantity.

Objection (2) already appears in Parmenides, B8. 9–10, which, like the present passage, seems to appeal to a tacit Principle of Sufficient Reason (cf. above on 251ᵃ17). If an event E happens at time *t*, there ought to be some reason why E happened at *t* rather than at some other time. Every event, then, should have some cause, and, in particular, this cause should give a scientifically adequate explanation of why the event happened when it did (cf. Barnes 1979*b*: i. 187–8).

252ᵃ17–22

The remainder of the paragraph makes it clear that the previous argument is directed not at Empedocles and company, whom the opening of the paragraph led us to expect as the target, but at those who do not have a cyclical account of change. The infinite periods of rest and motion characterize a cosmos in which the beginning of motion happens only once; the inexplicable event of incipient motion occurs in such a cosmos. If one is to have both cosmic rest and cosmic motion, it is better to envisage cyclical change, because at least here there will be some principle of order, e.g. the alternating dominance of Love and Strife in Empedocles' world.

252ᵃ19: 'it is better to say with Empedocles': i.e. than to say with Anaxagoras that the cosmos has a unique beginning.

252ᵃ22–7

Although a cyclical cosmos is better than one that is generated only once, Aristotle does not accept the cyclical account either. Here he complains that Empedocles never really explains why motion must be intermittent in the cycle. There are, indeed, two cosmic forces, but they directly account for different sorts of motion, not different periods of motion and rest (^{a}25). Some deduction is needed of their cosmic effect from their causal activity. Empedocles might reply that the explanation is implicit in the story of cyclical development: in this situation Love does this, Strife does that, because Love joins together, Strife separates, etc. Yet Aristotle would still want an argument to show exactly why each force acts as it does.

In general, Aristotle desires Empedocles to specify the domain in which Love and Strife properly apply. Of course, they apply primarily to the realm of human relationships—not to cosmology. To get to cosmology, one will have to make an inductive leap; Aristotle wants the leap to be made explicit and defended as an appropriate generalization. In effect, he is demanding a scientific method, not a speculative insight, however grand.

252^{a}27–32

252^{a}28: 'in turn' (*en merei*): see note on ^{a}8. In the human world Love and Strife (if indeed we can isolate these as objective powers) act concurrently. Why, then, should one of them dominate at one time and another at another time? Empedocles may be thinking of cycles of war and peace between city-states, or of political harmony and strife within the city-state as a model. But again, this kind of tacit analogy rates only as a beginning point of philosophical or scientific investigation for Aristotle, as indeed for us.

Elsewhere (*Met.* A4, 985^{a}21–9, B4, 1000^{a}24 ff.; *GC* II. 6, 333^{b}19–21) Aristotle criticizes Empedocles for not giving an adequate characterization of Love and Strife: although Love is supposed to join things, it separates the portions of the elements when it compounds them with other elements; and although Strife is supposed to separate things, it joins portions of the same elements. The criticism is not completely fair: on the political model Empedocles presupposes, the real challenge is to unite unlike bodies, e.g. rich and poor classes; Love joins them, Strife tears them apart. Cohesion within the class does not require explanation, but adhesion between classes does. Nevertheless, it remains true that Empedocles does not give anything like a scientific definition of the forces that rule the universe. Aristotle also complains that we need a further cause to account for the beginning of the activity of Love and

Strife, respectively, and this will be the real first cause (*GC* II. 6, 334ᵃ7–9).

252ᵃ31:　Not only the alternating dominance of Love and Strife, but also the equality of their respective dominions, needs an explanation.

<h2 style="text-align:center">252ᵃ32–ᵇ5</h2>

Aristotle launches into a diatribe about explanation of universal conditions. From Democritus, he extracts a claim or assumption that everlasting conditions need no explanation. Whether Democritus asserted or implied such a view is unclear. If Aristotle is merely inferring from the remark he cites at ᵃ35, his evidence is weak. Democritus could be simply invoking a principle of the uniformity of nature in reconstructing the prehistory of the cosmos. Ps.-Plutarch, *Stromateis* 7 (Diels 1879: 581. 8–11), says only that the chain of causes which predetermines (*prokatechesthai*) events past, present, and future has no beginning. It is also possible that Aristotle is citing Democritus merely as someone who appeals to phenomena that always happen (in whatever context) as some sort of explanation for them. Of course, if this is so, Democritus may not be trying to *explain* the phenomena, but just to convince us that whatever explanation he gives is consistent with experience. Hints in Aristotle (*Cael.* III. 2, 300ᵇ8–11; *GA* II. 6, 742ᵇ17ff.) suggest that Democritus appeals to the always specifically in accounting for the motion of the atoms. In any case, Aristotle is disappointed that his predecessors did not explain intermittent change on the basis of some prior principle. Aristotle himself will acknowledge the principle of uniformity of nature later (7, 261ᵇ24–6).

Barnes (1979*b*: ii. 129–30) maintains that Aristotle is ascribing to Democritus a regularity theory of causation, and criticizing it as inadequate. According to Democritus, to explain an event is just to show that it fits into a pattern of regular occurrences in which physical objects behave the same way every time. According to Aristotle, to locate a regularity is not to explain, but to provide the starting-point for a genuine explanation that will tell why it has to be so. Although we are not in a position to judge whether Aristotle's reading of Democritus is fair, this metatheoretical difference seems to separate Aristotle from his Presocratic forebears.

Aristotle seeks to render the cosmos maximally intelligible, whereas the atomists seem to settle for a description of regularities.

251ª35–252ᵇ2: According to Aristotle's theory of scientific explanation, first principles are knowable but not demonstrable, while other universal claims are subject to proof or disproof (*An. Post.* 1. 1–3). That the interior angles of a triangle are equal to 180° is an everlasting truth, but, contrary to what Democritus's view entails (if he holds the view Aristotle attributes to him), it is provable. The theorem seems to have been a well-known fact of geometry: the proof, according to Eudemus, goes back to the Pythagoreans (Proclus, *in Euclid.* I, 379. 2–6 Friedlein; a different, less self-contained but equally elegant, proof is given in Euclid 1, prop. 32).

252ᵇ3–4: 'being everlasting . . . everlasting' (*aïdiotēs . . . aïdios*): here one wants to draw a contrast between what is everlasting, i.e. exists at all times, and what is timeless, non-tensed, eternal, outside time, or in general time independent in any of several possible senses. Nevertheless, Aristotle seems not ever to countenance the timeless. It is controversial whether his predecessors do (see Owen 1966); but whatever moves were made in the direction of a timeless existence, the closest Aristotle comes is the omnitemporal. (See Hintikka 1973: ch. 4.) In order to avoid misleading associations with more exotic concepts of eternity, I translate *aïdios* and cognates by 'everlasting' and cognates (cf. Guthrie 1965: 29). As Hintikka notes (1973: 83–4), the term *aiōnios* was available for 'eternal', but is not used with that meaning by Plato or Aristotle.

252ᵇ4: 'cause' (*aition*): the Greek term is more general than the English term, at least in scientific contexts (in ordinary language the English term is almost as broad as the Greek: Graham 1987: 165). A cause supplies an answer to the question, 'Why?'; i.e. it supplies the basis of a 'because': *Phys.* II. 7, 198ª14–16; *An. Post.* II. 11, 94ª28ff.; *Met.* Z17, 1041ª9–28; Wicksteed and Cornford i. 126–7, 165; Wieland (1962: 261–2); Vlastos (1969: 292ff.); Hocutt (1974); Moravcsik (1974). (Note, however, that a cause is not a mere linguistic explanation, but a real entity that grounds the explanation: Moravcsik (1991). On Aristotle's development of the four-cause theory, see Graham 1987: ch. 6.) If the 'why' asks for an explanation in terms of a static principle, the answer will not be a 'cause' in the strong sense usually used in scientific accounts. For example, 'Why is the sum of angles of a triangle equal to 180°?' 'Because . . . [a proof follows].'

Concluding Remarks: The Structure of the Argument in Chapter 1

Although the general tenor of Aristotle's argument is clear enough, the precise structure of the argument is less clear. Aristotle begins his argument (250^{b}23) proper with a distinction between a cosmological model in which motion begins once and for all after an indefinite period of rest (here Anaxagoras provides the paradigm) and a model in which motion and rest alternate (as in Empedocles). This leads us to expect that his refutation of alternatives to the view that motion is everlasting will fall into two corresponding parts. And indeed, we can read the argument as exhibiting a twofold structure, with a criticism of Anaxagoras's model beginning at 251^{a}8 and a criticism of Empedocles' model beginning at 252^{a}5. Then the major points of the argument would be as follows (cf. Appendix I: Outline of the Argument):

(I) Motion has always existed in the world.
    (A) There are two ways in which motion might not always have existed (250^{b}23).
        (1) Motion might begin after an infinite period of rest.
        (2) Motion and rest might alternate.
    (B) (A1) is impossible (251^{a}8).
    (C) (A2) is impossible (252^{a}5).

One crucial question is what Aristotle thinks he has proved when he rules out intermittent motion at 252^{a}4. He seems to be ruling out the change from motion to rest, or vice versa, in general, not alternating cycles of motion and rest as in Empedocles (see note *ad loc.*). But if that is so, it seems he has not addressed one of the key models which his argument is to refute; so he might be expected to take that up in the final section of the chapter, i.e. section C, which should thus be concerned with Empedocles' position.

I am tempted to read the argument of the chapter as being developed in this way. But a close reading does not support the interpretation. Section C begins with a reference to those who defend the view just rejected on the grounds that it is natural ('And the same goes for saying that things are naturally thus': 252ª5–6). The view just rejected is (A1), so it seems that Aristotle has still not narrowed his criticism to (A2). Furthermore, Anaxagoras is brought into the discussion (ª10–11), as though Aristotle is not focusing on the contrast between Empedocles and him, but on the possibility that Anaxagoras, too, might use the defence that intermittent motion is natural.

Thus, while the reading suggested in the outline above has attractive features, and might well have provided a tighter structure to the argument of the chapter, it does not seem to represent Aristotle's argument. For the real structure of the argument see Appendix I.

CHAPTER 2

Here Aristotle sets up some possible objections to his own account of change as everlasting, and briefly responds to them. The objections provide both a justification for opposed theories and a source of puzzles or problems (*aporiai*) of the sort Aristotle commonly uses to test the power of his own explanations. His own theory should be able to expose the failures of opposing theories, perhaps by diagnosing the errors inherent in arguments on their behalf. It should also be able to resolve the problems. Indeed, the solution of a problem is a discovery (*heuresis*, *NE* VII. 3, 1146ᵇ7–8). For an influential discussion of Aristotle's dialectical method, see Owen 1961 (see further below, note on 3, 254ª30). Note, however, that Aristotle does not attribute any of the arguments to any particular philosopher. He seems to be reconstructing possible arguments for the theory that motion is not everlasting.

252ᵇ9–12

The first argument that motion is not everlasting presumes a background of Aristotelian assumptions. Aristotle individuates motions by their 'whence and whither'—i.e. by their starting-point and end-point—as well as by the subject of the motion, the cause of the motion, and the time of the motion (*Phys.* V. 1; see Penner 1970). Motion is defined in part by its beginning and end-points, which provide its limits; if there are no limits, the motion will be undefinable and unintelligible. Moreover, an infinite distance cannot be traversed (*Phys.* VI. 10, 241^{a}26–^{b}11). But in any case the universe is finite, so rectilinear motion must be finite (*Phys.* III. 5; *Cael.* I. 9). Aristotle himself endorses such arguments (cf. ^{b}28–31), but it is doubtful that any of his predecessors would. In particular, the atomists hold that the universe is infinite, and so would reject Aristotle's denial that infinite motion is possible.

252^{b}12–16

The second argument makes a connection between alternating motion and rest in inanimate objects and such motion in the cosmos. The fact that individual sensible objects are sometimes at rest and sometimes in motion is an indication of the way in which the world as a whole can alternate between rest and motion. If it is not possible for motion to come to be where it was not before, then individual objects should not be able to alternate between rest and motion; but if they do alternate, then it must be possible in general for motion to appear where it did not exist before.

252^{b}17–28

The third argument is similar to the second—indeed, we may view it as an extension of it—but it concentrates on the striking ability of living things to originate their own motion. Inanimate objects require some external agency (*dunamis*) to set them in motion, while animate objects have their own internal agency (*phusis*). Actually, this is not quite in line with Aristotle's view: he holds that simple bodies have internal agency with regard to their proper motion (*Cael.* I. 2; *Phys.* II. 1). Fire and air travel up to their natural places, earth and water travel down. The fifth element travels with circular motion. But apart from a primitive internal agency, inanimate objects do not originate their own motions. By contrast, living things are capable of complex and unpredictable motions; thus they provide the best examples of original or apparently uncaused motion for comparison with alleged original motions in the cosmos. Aristotle's arguments later in Chs. 4 and 6 will call into question his own established principles of natural motion. For now he presumes these principles as part of the objection.

Waterlow (1982: 220–1) points out that this objection is premature: nothing Aristotle has said commits him to the two claims presupposed by the objection: namely, (a) there is a single change predicated of the universe as a whole, and (b) this change is everlasting. So far, all Aristotle has committed himself to is the claim that there is always change in the universe. It is fully compatible with this claim that this change might reside in different bodies which come into motion and cease their motion at different times, so long as at least one of them is always in motion. Clearly Aristotle is anticipating his own stronger position and objections that could be raised to it.

This point suggests a further problem, not noticed by Waterlow: objection (1) is also premature. For, like objection (3), it assumes that there must be a single continuous motion that is everlasting. If one's view were that there is always motion in the universe, but that this motion consists of episodes of finite motion (perhaps by perishable subjects), (1) would pose no threat. Incidentally, the same account applies to the second objection too, so that all three objections presuppose a position that Aristotle has not yet articulated.

Aristotle already seems to be assuming that the subject of cosmic change is the cosmos itself, which for Aristotle is identical with the whole universe. Though nothing he has said so far entails this view, he is thinking of the universe as the subject of a unitary everlasting change, which must now be explained and justified.

252^b26-7: The microcosm/macrocosm distinction derives from Democritus B34. Many Presocratics saw our world as a closed cosmos inside an infinite expanse of continuous matter (Anaximander, Anaximenes, Anaxagoras) or an infinite expanse of unorganized atoms and void (Democritus). Aristotle adopts the picture for the sake of argument; but of course he rejects the notion that there is anything outside the finite cosmos bounded by the sphere of the stars. See *Cael.* I. 5–9, esp. 9, 279^a9ff., and *Phys.* III. 5, 204^a34ff. (It is true that he makes obscure remarks about 'the things there' (*ta'kei*), i.e. outside the heavens, enjoying the best kind of life: *Cael.* 279^a18-22; but in any case such things do not have body, $^a16-17$ cf. on 10, 267^b19-26.)

252^{b}28–253^{a}2

Not surprisingly, Aristotle accepts the principles enunciated in (1). However, later on in Ch. 8 he will show how circular motion has special properties which allow it to be everlasting. All of Aristotle's answers here are programmatic, and will require a full elucidation later in his treatise.

253^{a}2–7

Aristotle will stress the role of the external mover in circumventing this difficulty. His reply is developed in Ch. 4.

253^{a}7–21

Aristotle's preliminary reply to (3) is unexpected. He rejects the proffered description of the situation: the animal that was now at rest, now in motion, was not really completely motionless when at rest. There is some motion going on in the animal all the time, and this motion must be a response to the environment. Even though an animal seems to be completely still, e.g. in sleep, we must posit some unobserved motion to account for its later movements. (See Aristotle's further discussion of animal motion at 6, 259^{b}1–16.) What Aristotle says here is compatible with a behaviouristic position based on a mechanistic physics. Every observed behaviour is to be traced to some previous behaviour or some environmental stimulus; every motion is to be traced to some previous motion. Aristotle's defence of teleology in *Phys.* II would lead us to expect a further analysis in terms of ends or goals rather than in terms of antecedent motions and dispositions of matter. Will Aristotle ground animal motion in teleology? How will the teleology of *Phys.* VIII square with that of *Phys.* II?

In fact, there seem to be two different solutions to the problem suggested in Aristotle's remarks. The point could be that the animal is never completely at rest: some motion is always going on, such as a heartbeat. Thus, the origination of motion can in principle be explained in terms of a chain of involuntary motions building from an unobservable to an observable change. Or the point could be that the environment is always impinging upon the animal, and

that some perhaps unobserved change in the environment can
trigger an observable response in the animal. The two sorts of
explanation are not incompatible, but they are independent of one
another in the sense that one might apply while the other did not.
When Aristotle says, 'The animal itself is not reponsible for the
movement of this [continuously moving] part, but perhaps the
environment is' (ᵃ12–13), he seems to conflate continuous involuntary motions such as heartbeats with reflexive responses to the
environment. But surely we do not want to make a heartbeat
dependent on the environment in the same way a knee-jerk is
(although of course a stimulus in the environment could increase
the heartbeat). Later, Aristotle will allow both kinds of explanation, though he will focus more attention on the second.

253ᵃ18

For Aristotle's account of sleep, see below, 6, 259ᵇ12–13 with comment and *De Somno* 458ᵃ10–32.

CHAPTER 3

Aristotle begins the chapter with a taxonomy of all possible positions relative to the question of whether things are in motion or at
rest. He began Ch. 1 with a schematic discussion of the beginnings
of motion; but there his question was a more limited problem of
cosmogony: did motion have a beginning or not? Here Aristotle is
concerned not with the beginning of cosmic motion, but with
motion in general—which is a condition of there being a cosmos—
and how it is distributed among bodies. His scheme is an a priori
one, but it is not therefore without connections to historical positions. The view (1) that all things are always at rest is that of the
Eleatic school, first advanced by Parmenides. The view (2) that all
things are always in motion is attributed to Heraclitus by Plato
(*Crat.* 401d, 402a; *Tht.* 152e, 160d; *Phil.* 43a). The view (3) that
things both move and rest could be called the common-sense view,
but clearly Aristotle wants to make precise philosophical distinctions among possible versions of the general position. Although he
makes no particular attributions of versions to historical figures,
and although the scheme may be designed merely for theoretical

completeness, one can find possible instances of all three versions of (3). That (a) some things are always in motion, some always at rest, suggests Plato's distinction between changeable particulars in the sensible world and changeless Forms. (b) The alternation of all things between rest and motion is instantiated by Empedocles, for whom the elements alternate between a completely homogeneous 'Sphere' at rest and a cosmos in which the conflicting powers of Love and Strife battle for sovereignty. Finally, Aristotle himself will argue for the view (c) that some things are always at rest, some always in motion, and some alternate between rest and motion. The very status of (3c) as a compromise view may in his mind constitute a presumption in its favour.

253ᵃ32–ᵇ2

Aristotle challenges the Eleatic view head on with what seems to be a rather crude attack (contrast his more detailed and subtle arguments in *Phys.* I. 2–3): the view is incompatible with sense perception, and it undermines experience. This complaint is hardly news to the philosophical tradition. But Aristotle repeats his point several times in the present chapter, and it would be well to consider whether he does not have a substantial criticism to make. A philosophical theory aims at accounting for our experience in some sense. (Note: where I translate 'calls into question *the whole of experience*', ᵃ34, the Greek has only 'some whole'.) If in attempting to explain experience we produce a theory according to which none of our experience is valid even for evaluating the theory itself, we seem to have undermined the whole project of philosophical explanation. Our theory has defeated the purpose of proposing the theory: namely, to explain our experience. This is not to say that a philosophical theory could not provide some radical reinterpretation of experience or parts of experience; we might be willing, as Aristotle hints, to allow our theory to undermine the possibility of physical science. But we could not tolerate a complete rejection by the theory of all our experience.

This argument may still seem to beg the question against Parmenides, in so far as he actually argues against the validity of sense experience (B7. 3–5; cf. B6. 4–9). But note that Parmenides' criticism of the senses does not supply the main premiss in his argument against motion and change. The key point is the claim that what-is-not cannot be an object of knowledge or of reference (B2, B6. 1–2). In attacking the evidence of the senses, Parmenides makes the limited claim that no confused evidence of the senses can outweigh the evidence of reason—i.e. his claim that in some sense what-is-not is not an object of knowledge and reference. Parmenides' point, then, is not that one can never use one's senses as evidence of anything, but that one may not appeal to the senses to counter a logical truth (cf. Barnes 1979*b*: i. 297–8). It is a reply to a potential objection, not a key premiss in his basic argument. Accordingly, if one has independent grounds for rejecting Parmenides' prior argument about what-is-not—independent, that is, from the evidence of the senses—then one need not be troubled by Parmenides' criticism of the senses. Aristotle does have an independent logical-metaphysical argument against Parmenides' understanding of what-is-not, which he expounds in *Phys.* I. 8 (cf. his criticisms of monism in 1. 3). Thus Aristotle can dodge Parmenides' criticism of the senses in the context in which it is intended, while attacking Parmenides for proposing a theory which is incompatible with sense experience.

In a similar attack on Eleatic philosophy (specifically Melissus's theory) at *GC* I. 8, 325ª13–23, Aristotle points out the danger of ignoring sense experience in following out one's arguments to the bitter end. Not even a madman, he assures us, is so mad as to think that fire and ice are one. The practical demands of life keep us from actually trusting in any theory which is too remote from our experience. One is reminded of Descartes's determination to adhere to normal practices of life while entertaining his methodic doubt of experience.

253ᵇ2–5

Cf. *Phys.* I. 2, 184ᵇ25 ff. While we may readily grant that foundational questions about nature do not fall within the province of physics *per se*, Aristotle's complaint here seems singularly disingenuous. For Aristotle himself has introduced the foundational questions into this treatise dealing with foundational issues. If he wishes to consider all possible views about cosmic motion, he needs to give the Eleatics their due.

253^{b}7–9

Cf. *Phys.* II. 1, 192^{b}20–3.

253^{b}9–11

Less assured than Plato (see references in introductory remarks to this chapter), Aristotle cautiously takes Heraclitus as propounding a radical theory of flux (*Cael.* III. 1, 298^{b}29–33; cf. *Top.* I. 11, 104^{b}21–2), and presumably he has Heraclitus in mind in the present passage. (Simplicius more vaguely refers to the Heracliteans on the strength of Plato's *Cratylus*, also noting that Alexander referred to the atomists.) It is controversial whether Heraclitus advocates the strong view that all things are in constant flux (he says much less than this, e.g. in B88, B126, B76). Against the flux doctrine are Reinhardt (1916: esp. 206–7), Kirk (1951*a*, 1954), and Marcovich (1965: cols. 289, 293); for it are Vlastos (1955), Popper (1959), Guthrie (1962: ch. 7), Stokes (1971: ch. 4), and Barnes (1979*b*: i, ch. 4). For a synthesis, see Graham (1997). Heraclitus's follower Cratylus, whose lectures Plato heard, did apparently hold such a view (*Met.* Γ5, 1010^{a}7–15—though some caution is in order in taking even Cratylus as a radical Heraclitean: Kirk 1951*b*, Allan 1954).

Here it may be helpful to distinguish between two different versions of the flux doctrine. Let us call the view that all objects are changing in *some* respect at every moment 'weak flux', and the view that all objects are changing in every respect at every moment 'strong flux', following Plato's distinction at *Tht.* 182c–d. Plato further suggests the possibility that even acts of perception themselves may be changing to something else (ibid. 182d–e). Plato uses increasingly radical interpretations of the flux doctrine to undermine the claim that knowledge is perception (182e–183b). But one can still consider the claims of flux in the context of the metaphysics of change itself, as Aristotle does here. In any case, Aristotle seems to grant for the sake of argument the possibility that either the weak or the strong flux doctrine may in fact be true. Minimally, the flux doctrine, in its weak version, maintains that there is continuous change of some kind in any subject. Hence Aristotle argues in turn against continuous change in several different categories:

increase and decrease ($^b13-23$), alteration ($^b23-31$), motion in place ($^b31-254^a1$), and finally (and more indirectly) coming to be and perishing (254^a10-14). On Heraclitus, see further below, on 8, 265^a2-7.

253^b11

'this escapes our senses': Kirk (1951b: 241) objects to Heidel (1906: 350ff.), 'It is most unlikely that Heraclitus ever held such a view', citing the present passage as the origin of a mistaken interpretation of Heraclitus. Kirk points to Heraclitus's approval of sense perception (cf. B55, B105a), and traces the view to Melissus B8. Kirk is, I believe, right; in his favour I would stress the lack of any interest by Heraclitus or his contemporaries in the microstructure of things. It is only after Parmenides that pores, effluences, and corpuscles become crucial elements of physical structure. (Alcmaeon had earlier posited the existence of pores, but only as conduits for sensations: Theophrastus, *De sensu* 26 = DK 24A5.)

In fact, even if one holds only to weak flux, one would be forced to say that ongoing changes escape our perception, since we do not perceive constant change in many things. Note, however, that it is possible that Heraclitus was concerned with changes of the great world masses (earth, sea, etc.), and hence held an even weaker view than weak flux: namely, that some change is always going on in the world masses.

253^b13-14

What Aristotle must mean here is not simply that there is some intermediate state between increasing and decreasing (as Ross and Cornford have it), but that there are periods in either the process of increase or the process of decrease in which no physical change is going on (cf. Wagner). For the following cases are meant to illuminate the present remark.

253^b14-23

The argument here follows a similar argument at *Phys.* VII. 5, 250^a9ff., which Aristotle may be alluding to at $^b18$. Aristotle's general point is that the defender of flux commits the fallacy of division. We cannot argue from the fact that the subject is infinitely divisible to the fact that its changes come about in continuously divisible steps. His point is quite correct, but his example is a bit puzzling. Why should Aristotle focus on the effects of the drops being cumulative rather than continuous? The very fact that we are dealing with discrete drops rules out the possibility that there is a continuous flow of water over the surface; why not just focus on the discontinuity between the drops themselves? As it is, Aristotle merely asserts that the individual drop has a cumulative rather than a continuous effect, and depends on the analogy with boat hauling to convince us.

253^{b}23–30

Aristotle goes on to make the stronger point that we have positive evidence against continuous alteration: freezing can happen suddenly, although presumably the period of cooling off takes time. Modern physics could support his point: although heat loss from a liquid may be continuous, it is only when an amount of heat equal to the 'heat of fusion' is lost that the liquid freezes to the solid state, and the change is an immediate one. For some time before the change the actual temperature of the liquid does not change; having reached the freezing-point, the liquid must give up its latent heat before the change of state takes place. Thus the observer perceives no change for some time, and then a sudden event of freezing.

Aristotle's next example of being healed is puzzling, for he stresses the seemingly opposite point that healing takes time. But perhaps he means to present the converse of the freezing example: in freezing a process taking time can bring about a sudden change of quality; in healing a sudden change of quality (from sickness to health) takes time to bring about. In both cases we must distinguish between an extended process of change and a sudden manifestation of the change. Note that the verbs 'heal' and 'cure' are what Ryle (1949: 130, 149) called 'success words' which describe 'achievements', or (in terms of Vendler's more technical terminology (1967: ch. 4), but not in Ryle's looser terminology) immediate changes. While convalescing or being treated may constitute extended processes, being cured constitutes an immediate change of state from sickness to health. In Aristotle's example we can say that the alteration to health happened suddenly as the result of a prolonged process. In this case it might be best to say that we are dealing with different events having different identity conditions. The distinction between the extended process and the sudden alteration might, however, provide ammunition for the defender of flux: the fact that the sudden change takes time to bring about must indicate that some change is transpiring even when we do not perceive it. For instance, water may be losing heat to the environment over a long period, while the actual freezing occurs only at the end of the period. There is an actual physical change going on even when it is not manifest to sense perception.

Aristotle may mean the healing example differently. His point may simply be that even if some alterations take time, there are still constraints on the change: there is a goal of the change. Hence the change cannot be to any chance state, but only to the opposite of sickness (cf. *Phys.* I. 5, 188^{a}31–^{b}8). Nor can change go on indefinitely, for it has a fixed end-point—in the present case, health. This interpretation gains some support from the reference to alteration between opposites at 253^{b}30–1.

253^{b}31–254^{a}1

Turning to motion in place, Aristotle points out how unlikely it would be for us to be unaware of whether a stone is in motion or at rest. The strong flux theorist must hold that a stone is always in motion, even when it is apparently at rest. Today it is easy to argue for constant motion, relative to appropriate frames of reference: a stone on Earth is moving about the circumference of the Earth with the planet's daily rotation; it is moving about the Sun with the planet's annual orbit; and it is moving about the centre of our galaxy with the motion of the solar system. But from his cosmically privileged frame of reference on a motionless Earth, Aristotle can appeal to sense perception to establish the fact that the stone is motionless. Today we can also point out that constant motion occurs at the atomic and sub-atomic levels. But in Aristotle's time there was no empirical evidence for such a claim, and in fact, Democritus had to make an appeal beyond any possible sense experience to support his view of atomic motion (B6–11). Thus Aristotle is on epistemological high ground when he insists that sense experience supports his view.

Aristotle's argument that the stone is not in motion will not by itself overthrow the weak flux doctrine; but in conjunction with his denials of constant change in other categories, it could count also against weak flux.

254^a9-10

In arguing that (3a) is impossible, Aristotle gives two absurd results—that there will be no growth and no forced motion—but only one reason, and that reason applies only to the latter result. He must have in mind a parallel reason for (3a) ruling out growth: without some fixed starting-point (some previous state of rest), there could be no growth. But this argument fails, for while we do indeed need to assume some fixed point with reference to which we measure, this point need not occur in a state of rest. For example, a plant may be growing from the time the stem emerges from the seed; we can point to the growth from time t_1 to time t_2 by comparing the size of the plant at the two respective times. But we need not suppose that there is any stretch of time at which the plant is not growing: our point of reference t_1 may simply be an arbitrary instant in a period of continuous growth. If Aristotle objects that the seed was in a state of rest before it was planted, we may reply that it is not the seed which is growing, but the plant itself; and the plant's existence may be dated from the beginning of the process of growth.

254^a11-12

'But it is virtually the general view that motion is a kind of coming to be and perishing': Aristotle has not yet discussed coming to be and perishing, which he sometimes treats as species of *kinēsis*, though more strictly speaking he classes them as species of *metabolē* ('change') as opposed to *kinēsis*, which comprises alteration, increase and decrease, and motion in place (*Phys.* V. 1). Curiously, he now treats coming to be and perishing as if they were corollaries of motion—on the grounds that motions involve the coming to be or the coming to occupy something. But properly, no one species of change is reducible to any other species of change for Aristotle, for each species of change is defined by reference to its own distinct category, and the categories are exclusive of each other.

Aristotle's point here is no doubt a dialectical one—based on what (other) people say about coming to be and perishing—but it still seems to miss the mark. The sense of 'come to be' in which it is true to say that all change presupposes coming to be is a very weak one, and one from which it does not follow that coming to be in the important metaphysical sense is done away with. For Aristotle's acute analysis of different sense of 'come to be' see *Phys* I. 7–8.

254ᵃ25–30

Melissus holds that the world is unbounded (*apeiron*) and motionless. In arguing against Melissus, Aristotle is not simply reiterating the point that we can see some things in motion. The point is that even if Melissus were right, we would *seem* to perceive motion, and this fact is sufficient to prove that there *is* motion. The argument may be set up as follows:

(1) Suppose that there is no motion.
(2) There seems to be motion (fact).
(3) If (2), then there is opinion (even if it is false), or there is imagination.
(4) Opinion is motion.
(5) Imagination is motion.
(6) Hence, there is motion (2–5).

The argument cleverly moves from the Eleatic's claim that motion only seems to exist to the conclusion that there is motion. Yet the Eleatic could resist the argument at two crucial points. First, he could try to object to Aristotle's attempt to reify opinion and imagination. Suffering an illusion is not some real existence, and hence one should not make an ontological commitment to it. Second, there is no reason to accept the classification embodied in (4) and (5) that these mental states are motions. Aristotle indeed claims that imagination is a motion at *An.* III. 3, 428ᵇ11, but there is no reason for an Eleatic opponent to assent to this claim. In fact, if the Eleatic is a monist, as Melissus seems to be, he should refuse to allow Aristotle to construe opinion and imagination as items distinct from Reality as a whole.

But Aristotle's argument is not hopeless. He is right to call attention to the Eleatic's appeal to seeming or appearance (prominent in Parmenides' poem). The Eleatic typically says that perception of change is mere appearance. But if there is appearance, it must constitute a mental state different from knowledge, and hence must exist in some sense (for if my opinion that there is motion is not different from knowledge, then I know that there is motion). The Eleatic could still argue that although in some sense there are different mental states of opinion and knowledge, yet at the ultimate level of description there is only what-is, which comprises all things. This move, however, would at least presuppose a kind of two-level analysis which would afford some limited existence to appearance. And since appearances themselves change, there must in *some* sense be change in this modified Eleatic world. Here we must note a possible ambiguity: we may have a belief *in* change or our beliefs may themselves change. It is not clear which line Aristotle wishes to pursue. But perhaps it does not matter much: if appearances change, my beliefs will change; and if my beliefs change, it is rational for me to believe in change. In either case, there are grounds, and indeed grounds suggested by the Eleatic himself, for thinking that change is an integral part of the world. In Alexander's succinct statement: 'If no entity moves and there is no motion, but some things appear to move, there is appearance (*phantasia*). If there is appearance, there is motion and some entities move. Hence if no entity moves and there is no motion, there is motion' (ap. Simpl. 1205. 2–5). Thus, even if the Eleatic rejected Aristotle's classifications in (4) and (5), he would have to grant that there are changing states of affairs, even if these are only mental states.

For Aristotle the *phainomena*, including (but not limited to) sense perceptions, provide the evidence for scientific explanation (see Owen 1961; Nussbaum 1982; Irwin 1988). An important component of the *phainomena* are *endoxa*, or authoritative opinions held by everyone, or by the majority, or by the wise, or by some sub-set of the wise (*Top* I. 1, 100ᵇ21–9). Any theory which begins by denying the evidence, including human consensus, will be self-stultifying. If we deny the source of our evidence about the world from the outset, how will we ever explain the world? In Aristotle's method, experience—taken in broad sense to include common-sense judgements as well as sense perceptions—is the beginning and end of scientific justification. We must begin from our experience to identify the facts and problems to be explained; we must explain them on the basis of arguments from true and (ideally) necessary premisses, and we must then show how the solution solves the problems and accords with the facts—i.e. we must show how experience vindicates our account. There is some room for debate about how well Aristotle's scientific method and his episte-mology go together: will evidence of the sort Aristotle favours support a foundationalist epistemology of the sort he seems to envisage in the *Posterior Analytics*? (See Irwin 1988 for extended treatment.) However this problem is resolved, one may grant to Aristotle the basic point that our beliefs must be revised in light of our experience in general and our sensory evidence in particular.

The reference to the better/worse, plausible/implausible, and fundamental/not fundamental isolates basic distinctions in ethics, epistemology, and metaphysics. Unless one has some appreciation of the hierarchical relationship between evidence and explanatory principles, one will never succeed in explaining anything.

254ᵇ6

To judge from the concluding remark of this chapter, we should next expect a discussion establishing (3c) and eliminating (3b). In fact, the next chapter has no obvious connection to the debate between the two remaining alternatives, being focused solely on questions of causation concerning things that can vary between change and rest—i.e. the class of beings which both alternatives recognize. Only in Ch. 5 will we begin to get evidence of an entity invariably at rest—when we meet the unmoved mover—and only in Ch. 6 will we get evidence of an entity invariably in motion—when we meet the outermost cosmic sphere. In Ch. 6 Aristotle does recapitulate his argument (259ᵃ20ff.) including the present

classification, but only after he has long ceased to follow the programme set up in Ch. 3.

CHAPTER 4

In this chapter Aristotle argues to the conclusion that every object
in motion is moved by something. The conclusion is, as we shall see,
surprising in light of Aristotle's theory of nature. Aristotle begins
by making some preliminary distinctions.

254ᵇ7–12

Aristotle makes a well-known distinction between incidental (or
'accidental' (*kata sumbebēkos*)) and intrinsic (or 'essential' (*kath'
hauto*)) motions. One's head may move in virtue of being on one's
body, which is walking; nevertheless, the real subject of motion is
the body, not the head. Thus, if we are to investigate motion
scientifically, we shall study the motion belonging to the subject
proper, rather than to some attribute or part that moves with the
subject. The point seems merely preliminary, but we shall see that
Aristotle seems to get into trouble precisely through the confusion
of incidental and intrinsic motions later in the chapter.

254ᵇ12–24

In introducing a division of intrinsic motions into self-caused versus
caused by another and natural versus forced, Aristotle does not
make it clear what precisely the relation between the two distinctions is. He seems to leave open the possibility that the distinctions
are completely independent of one another. Hence we could set
them on different axes and see if all possible combinations are
realized (see table).

	moved by nature	moved by force
moved by self	1	2
moved by another	3	4

In the following examples, we have (1) a case of an object moved by self and by nature: namely, an animal moving itself naturally. Objects moved by force, such as earth moving upward, are moved by another, and thus belong to case (4). If these were the only possibilities, the two distinctions might collapse into one another, with moved by self and moved by nature being coextensive, and moved by another and moved by force being coextensive. But we can supply another possibility from other texts: (2) that of a self-moved object being moved by force, e.g. in the case of a doctor healing himself (*Phys.* II. 1, 192^{b}23–7; 8, 199^{b}31). Yet Aristotle seems to say at ^{b}14 that all cases of self-causation are cases of natural motion. The healing action in our example is not natural, but it is brought about by the same individual who is healed. Why does Aristotle not consider this example? At *Phys.* II. 1, 192^{b}23–7, he points out that it is a coincidence that the agent and the patient are identical. This of course helps him explicate his sense of 'nature' whereby the source of change is in something not in an incidental way. But we have already conceded that this change is not *natural*. What we wish to know is why it is not a legitimate case of *self-motion*. Aristotle does seem to hold to a strict sense of 'self-caused' here, one which may apply only to non-voluntary motions. But he does not defend or explicate his sense of self-causation. What, then, about the last possibility: (3) being moved by nature and also by another? The remainder of the chapter is chiefly about this case.

254^{b}24–33

Here Aristotle announces, prematurely, the thesis of the chapter: everything in motion is moved by something. The thesis seems immediately problematic in the case of objects that move by nature. One can say, as Aristotle does here, that animals move themselves, and hence that they are moved by something. But what of other natural movers such as simple bodies, and even plants? Do they cause themselves to move in any important sense, or do they simply move?

Here it will be helpful to make some distinctions of our own (borrowed in part from Waterlow 1982: 39 ff., 162 ff.). Let us define *intransitive* motion as motion in which object A simply moves without being caused to move by anything else. We can describe an intransitive mover A as moving *on its own*, or *originating* its own motion. Intransitive motion will be properly described in a sentence of the form 'A moves', where the sentence cannot be further analysed into 'A moves B' or 'B moves A' or 'A moves A'. Let us define *transitive* motion as motion in which object A moves object B. In this case we clearly have a mover and a moved, an agent and a patient. Let us further define *reflexive* motion as motion in which A moves A. Reflexive motion can be viewed as a special case of transitive motion in which the agent and the patient coincide. Now relative to these distinctions, Aristotle seems to be ruling out the possibility of intransitive motion in favour of transitive and reflexive motion. But this seems surprising in light of his original discussion of natural motion at *Phys.* II. 1. There Aristotle described a nature as an internal source and cause of motion or rest. By making the source and cause internal to the natural object, he seemed to suggest precisely that natural objects move by themselves, or move intransitively. For although Aristotle did not formally distinguish between intransitive and reflexive motions (but note 192^{b}23–7), he clearly did not conceive of the nature as acting on anything in the natural body, i.e. as being an agent with a patient. Here Aristotle will reject the possibility of natural motion being intransitive motion.

The above distinctions help explicate a distinction of Aristotle's own making, that between nature and *power*. Whereas nature is an internal source of change, power is an external source. Specifically, it is 'the principle of motion or change in the other *qua* other' (*Met.* Δ12, 1019^{a}15–16, Θ1, 1046^{a}10–11; *Met.* Δ12 gives multiple senses of *dunamis*, but it does not define the broadest sense of the term, potentiality; the sense it does define I shall translate as 'power'). There are active and passive powers, corresponding to the ability to change something else and the ability to be changed by something else, respectively. Aristotle uses crafts such as building and medicine as paradigm examples of a power. The builder has in himself a principle of changing bricks and boards into a house. The crucial difference between the growth of a tree and the building of a house is that in the former case a seed has an internal source of change that is able to transform materials of the environment into organic parts of a tree, while in the latter case an external agent must take appropriate materials and organize them into the parts of a house.

The general outcome is the same for both; indeed, if houses had natures, they would produce the same results by the same steps as builders now accomplish them (cf. *Phys.* II. 8, 199^{a}12–13, ^{b}28–9). In terms of our previous distinctions, nature produces intransitive motions, whereas power produces transitive motions. Reflexive motions, such as that of a doctor healing himself, are special cases of transitive motions in which the agent and the patient in a transitive motion coincide.

Powers are exhibited in nature as well as in art. For instance, fire has the natural power to heat objects. Nevertheless, Aristotle tends to associate power with non-natural changes and nature with natural changes. In *Phys.* II. 1 Aristotle's main contrast is between natural changes and artificial changes. And at *Met.* Λ3, 1070^{a}6–9, Aristotle identifies art as a principle of change in another, contrasting it with nature as a general type of causality. Thus he seems to have the tendency to read the nature–power contrast as a nature–art contrast, which in turn amounts to a contrast between non-deliberative and deliberative teleology (cf. *Phys.* II. 8, 199^{a}8–15, ^{b}26–32).

The argument that emerges from the present paragraph of the text is as follows:

ARGUMENT A

(1) Of intrinsic motions, some are caused by the object itself, others by external objects ('by another').
(2) What is moved by the object itself is moved by nature.
(3) What is moved by another object is moved either by nature or contrary to nature.
(4) What is moved by another object contrary to nature is obviously moved by something.
(5) What is moved by itself is moved by something.

Aristotle is using the distinction between self-caused and other-caused motion as the main distinction to drive his enquiry. But this approach gets him into trouble. For if the disjunction in (A1) is exhaustive, then that premiss by itself already entails that all motions whatsoever are caused by 'something'. But this claim is highly problematic in the case of the natural motions of simple bodies, as we have seen. The distinction between objects moved by themselves and those moved by another seems to preclude the possibility that an object may move *by itself* without being *moved* by itself. Aristotle's appeal to objects moved by another could further the argument in (A4), where it follows analytically from the fact that an object is moved by another that it is moved by something. But his statement of the point (254ᵇ23–6) seems to stress the phenomenology of events contrary to nature rather than the analytical character of the assertion. Thus the appeal to a distinction between objects moved by themselves and those moved by another is introduced without sufficient support or motivation, and it seems to beg the question.

Indeed, if we take Aristotle's scheme as a classification, the way it seems to be intended, we immediately see its flaws.

   I. Incidental motions
  II. Intrinsic motions
     A. Moved by self
        1. Moved by nature (e.g. animal)
     B. Moved by another
        1. Moved by nature (e.g. earth moving down)
        2. Moved by force (e.g. earth moving up)

If our case of the self-healing doctor can be applied, motions caused by force will also be found in both IIA and IIB. Tacit in both IIA and IIB is the notion 'moved by something', which leads us to ask whether there might not be a higher genus above A and B but below II, in which II bifurcates into 'moved by something' and 'not moved by something'. (Aristotle criticizes the bifurcation into positive vs. privative or property vs. its contradictory divisions in biology, *PA* II. 3, but here, where we are dealing with more abstract possibilities, we cannot rule out the contradictory differentia a priori.) In any case, it is simply not clear that 'moved by itself' and 'moved by another' exhaust the possibilities.

It will not do, then, for Aristotle to give a hasty a priori argument for a highly questionable conclusion. He must back off and provide an analysis of natural motion—as he will now proceed to do. But had he started out with the distinction between natural and non-natural motion instead of the distinction between motion caused by self and motion caused by another, he could have provided a continuous argument, rather than one which he must interrupt halfway through to make his conclusion plausible.

Aristotle could have arranged his argument as follows:

ARGUMENT B

(1) Of intrinsic motions, some are natural, some are forced.

(2) What is moved by force is moved by another.

(3) What is moved by nature is moved either by itself or by another.

(4) What is moved by itself is moved by something.

(5) What is moved by another is moved by something.

Aristotle would then be able to draw his desired conclusion: that all bodies in motion are moved by something. At this point (B3) becomes the problematic premiss, in so far as Aristotle seems determined to exclude intransitive motion. But at least this version allows us to focus attention on the important issue, the analysis of natural motion, rather than, as in (A3) above, obscuring that issue.

254^{b}33–255^{a}5

Aristotle rightly notes that the most difficult point for his argument is making sense of the claim that non-reflexive natural motion is caused by something. His attempt to explicate the claim will take up the remainder of the chapter. Aristotle focuses on one case which is a paradigm case of natural motion, but which is not a case of reflexive motion: the motion of the heavy and the light. Heavy things go down by nature, light things go up by nature. There are two heavy elements, earth and water, and two light elements, air and fire. But there seems to be no agency, internal or external, to which one can refer their motion: they simply go up or down according to their nature.

255^{a}5–20

The heavy and the light move by themselves, but they are not *moved* by themselves. Thus it is difficult to say what agency causes them to move. Aristotle gives two arguments for the claim that they are not moved by themselves.

(i) If they moved themselves, they should be able to cause themselves not to move or to move in the opposite direction. Here Aristotle seems to equate self-motion with the possession of a rational potency—the ability to act or not act at will (*Met. Θ*2).

Rational potencies are found only in the soul, and they seem to be coextensive with the domain of voluntary action (ibid. 5, 1048ᵃ10–11). But why does Aristotle make self-motion coextensive with rational potencies? Why should we not say, for instance, that plants cause themselves to move? They seem to have the requisite complexity (see argument (ii) below) and to manifest flexible responses to environmental situations. Plants do have complex capacities in virtue of which they are alive (and hence they have souls, according to Aristotle), yet they lack the rational or quasi-rational functions that govern animal motion *per se*. Aristotle perhaps wishes to leave the door open to plant development as an instance of self-motion (ᵃ7: 'animate things', i.e. things with soul); but it is not clear how restrictive his notion of a power of contrary change is, and consequently whether plants could qualify as self-movers. For instance, a sapling normally ('naturally') grows upward toward the sun. But if it were in a position, say under an outcrop of rock, where it could not reach the sunlight by growing upward, it would grow sideways or downward. The sapling would thus be able to act contrary to its natural tendency. But it seems false to say that it could have stopped its motion—i.e. its phototropic growth. Would its growth, then, count as a case of self-caused motion? In some sense it surely should. But Aristotle's example of walking or not walking seems to presuppose a more complex function of soul: the sensitive soul driven by appetite. In any case, Aristotle will shortly find that animal motion, too, has a source outside the animal (6, 259ᵇ7 ff.). But we are left uncertain as to Aristotle's precise criterion for self-change. Aristotle further weakens his argument by a *non sequitur* when he goes from his claim that self-movers can stop their own motion (ᵃ7–9) to his conclusion that fire should be able to move down as well as up (ᵃ9–10): stopping an action and performing the contrary action are not the same thing. For example, a plant may be able to stop growing under certain conditions (e.g. a deciduous tree in winter) without being able to shrink. While the performance of either action might be a sufficient condition to establish self-motion, the failure to perform the latter would not necessarily count as evidence against self-motion.

(ii) Aristotle next argues (ᵃ12 ff.) that something continuous and uniform cannot move itself. The four 'elements' are simple bodies which are continuous and of a single, undifferentiated stuff. But such a substance cannot move itself. For in order to move itself, it would have to be articulated into at least two distinct parts, one which acts and one which is acted upon. But the elements have no distinct parts. This argument does seem to work, given Aristotelian assumptions about simple substances. To attribute reflexive motion to a substance, we would have to identify at least notionally distinct parts; but there can not be any. (Not that this point does not follow immediately from the fact that the substances in question are 'simple'. For 'simple', as Aristotle applies the term to primary bodies, means having a simple motion (*Cael.* I. 2, 268ᵇ27–8). The simple substances can be analysed into a complex of two powers and prime matter (*GC* II. 1–3); e.g. water is wet and cold matter. But these simple (in motion) substances are as materially simple as any substance can be, so we are justified in saying they have no distinct parts.)

Aristotle does not say precisely what parts an object needs to support reflexive motion. Possibly it needs a soul and a body, and the body is moved *by* the soul (this seems to Simplicius, 1208. 30–1, to be the obvious interpretation of animal motion). If this is right, the argument from rational potency would underline the need for a soul, which could act to produce either a given effect or its opposite. In order to have a soul, a body needs to have organs (*An.* II. 1, 412ᵃ27–ᵇ6); hence any physical body would at least need to have organs to be a self-mover. But since organs are an-homoiomerous or non-homogeneous, the body would have to have a relatively high level of composition. Aristotle's reference to an animate thing moving inanimate things seems to support this interpretation: the soul moves the body just as the animate moves the inanimate.

255ᵃ20–30

A transitive motion may be either natural or non-natural. For instance, fire heats objects naturally, whereas a lever raises objects non-naturally. The difference may lie in how the object of the action changes: it is unnatural for a heavy object to go up. Wood, on the other hand, would become heated because it is 'heatable' (*thermanton*), and it meets a heating agent (*thermantikon*); it could even be destroyed by burning, because it is 'burnable' (*kauston*) and it meets a burning agent (*kaustikon*). But we could explain the distinction between natural and non-natural transitive motions not only in terms of the output, but also in terms of the input. The lever requires a factor in addition to the lever itself: namely, a mover to push down on the lever, while fire simply heats everything in its environs. Fire, in other words, is active in and of itself.

Thus far, Aristotle has proved that the motion of the heavy and the light is not reflexive motion. But it remains mysterious just what kind of motion it is. Aristotle again appeals to the notion of potency to explicate the problem. But whereas he had previously invoked rational potencies to show what kind of motion the motion of simple bodies was *not*, he must now invoke a different sense of potency to say what kind of motion simple motion *is*.

He introduces a distinction in kinds of potency which is familiar from other texts (e.g. *Protr.* fr. 14 Ross; *An.* II. 1, 412ª22–6, 5, 417ª21 ff.). When I was a young child, I was potentially a reader in the sense that if someone taught me, I would be able to read words. But after a process of schooling, I learned to actually read words. Then I was a potential reader in a different sense: if someone put words in front of me, I could read them. We may distinguish these senses by saying that I was a potential$_1$ reader before I attended school and a potential$_2$ reader afterwards. Furthermore, I am an actual$_1$ reader after I attend school when I am not reading, but an actual$_2$ reader when I am reading words (cf. Kosman 1969). The transition from being a potential$_1$ to a potential$_2$ reader is a lengthy one; the transition from being a potential$_2$/actual$_1$ to being an actual$_2$ reader is instantaneous. I simply apply my skill on being presented with some appropriately inscribed material.

Aristotle draws an analogy between the simple bodies and his well-known cases of multiple potency. The heavy—say, water—is potentially$_1$ light; we heat it, and it turns to air; now it is actually$_1$ or potentially$_2$ light, for it has a tendency to go upward. But if it is impeded from going up, it is not yet actually$_2$ light. The air will be fully actual when it is at rest in the proper place of light things, in the upper regions of the sublunary world.

255^{b}13–17

Why does the light move up? Because it is the nature of the light to be up. This answer is fully in accord with Aristotle's account of natures in *Phys*. II. 1. But it seems to preclude the kind of answer he is looking for here. For if the light goes up simply to be up, then there is no need to ask further questions about what caused it to go up. To appeal to a nature is to put an end to a series of questions, to provide closure to an enquiry. Yet Aristotle needs another answer to satisfy his claim that everything that is in motion is moved by something.

255^{b}17–29

Aristotle now appeals to his multiple sense of potency to explain how a moving cause is still needed. He identifies the need in the gap between being potentially$_2$ and actually$_2$ light. The gap in this case must be supplied by some obstacle, or the potentially light would immediately become actually light. If there is some cause of the change from potentiality$_2$ to actuality$_2$, it must be identified with the agent who removes the obstacle.

But the agent does not really *cause* the light to move up; it simply removes the obstacle that keeps the light from moving up by its own agency. To see this, we need only imagine what would happen when I remove a lid from a pot of heated air if the air were *not* light: my action would accomplish nothing. My action would not *make* the air go up, but would only *allow* it to do what it was naturally inclined to do. Thus the external agent is only an incidental cause of motion, as Aristotle recognizes at ^{b}27. Yet at the beginning of the chapter, Aristotle dismissed incidental motions as irrelevant, and rightly narrowed his discussion to cases of intrinsic motion. If the agent removing the obstacle is only an incidental cause of the motion, it may safely be left out of account in specifying the proper causes of the change.

Is the cause of motion then the agent or agency that causes the potentially$_1$ light to be potentially$_2$ light—i.e. makes water become air? Consider the parallel case of learning. Does my primary school teacher, who taught me to read—i.e. led me from being a potential$_1$ reader to being a potential$_2$ reader—cause me to understand the newspaper I read today? Yes and no. If you are asking how it came about that I learned to read, I shall mention my teacher. But if you are asking what is the present, proximate cause of my reading—i.e. if you are asking a systematic rather than a genetic question—I shall not mention her. I read the newspaper because I have the potentiality₂ for reading and a newspaper to exercise it on. My teacher is responsible for my originally acquiring this potentiality, but she plays no role in my present actualizing of it. Thus, to invoke my teacher in explaining scientifically why I am reading the newspaper is to confuse two very different questions. In the same way, to invoke the producer of air in explaining in general why air goes up is to confuse a question about the genesis of a portion of air with the question of why air, whatever its origin, goes up.

There is, then, no *proper* sense in which the actuality₂ of air is caused by something. As Aristotle said at the beginning of the paragraph, the light goes up because it is the essence of the light to be up. We can multiply contributing causes, but they remain incidental to the upward motion itself. Nothing in Aristotle's argument undermines the impression that the simple motion of the elemental bodies is intransitive motion.

255ᵇ29-31

In rejecting the claim that simple natural motion is reflexive motion, Aristotle hits on the solution that the simple bodies exhibit passive transitive motion. Properly we should not say, e.g., 'air moves up', but 'air is moved up (by something)'. Thus air is moved up by the heating agent making the water into air, or by the removing agent taking off the lid of the boiling pot. But this analysis is surely wrong. The heating agent acts on water, not air; and the removing agent moves the lid, not the air. As we have already noted, if air did not move up naturally, no amount of heating water would make it do so, and no taking off lids would make it do so. The heating agent may change water into air, but it does not originate upward motion of the air; the remover of the lid may allow the air to rise, but it does not make the air rise.

Richard Sorabji (1988: 220) tries to establish the creator of the simple body and the remover of obstacles as 'low-grade' causes and to find a place somewhere between full-blown efficient causes and

84

accidental causes (p. 222). But there is no room here for a *tertium quid*. If the 'low-grade causes' are not real efficient causes, but rather, as they seem to be, only concomitant causes or necessary conditions, to recognize them will not provide evidence that everything in motion is moved by something.

In an interesting study, Helen Lang (1984) explicates Aristotle's theory by appealing to an essential cause as responsible for the change:

The elements are primarily moved by their essential cause or actuality, e.g., 'upward' moves water from the pure potential to be light to the habit of being light which is air; this habit will be immediately actualized unless something hinders. A hindrance requires an accidental cause to remove it and in this limited sense to initiate the motion. (p. 99, my emphasis)

While 255ᵇ15–17 might seem to support this reading, the italicized phrase corresponds to nothing in the text. The element surely actualizes itself, but Aristotle never says here or elsewhere that the elements are *moved by*, much less *primarily moved by*, their actuality. To say this would, I believe, be a category mistake: an essential cause cannot satisfy the phrase 'moved by ——'. For example, the sentences 'Socrates was moved by his humanity', 'The horse was moved by its equinity', are not simply false, but ill-formed; the point is not that they sound odd, but that they confound formal causes with efficient causes. (Formal and efficient causes sometimes coincide, e.g. when the parent's form is a moving cause; but that coincidence is ruled out in the present case.) Similarly, 'The air was moved by its lightness' would be ill-formed in Aristotelian logical grammar (though the sentence 'The air moved up *because* it was light' would be both well-formed and true). When Aristotle recapitulates his argument at ᵇ31 ff., the only candidate he offers for the causal agent is the creator or the remover of obstacles (see Graham 1996). Aristotle's appeal to nature at ᵇ15 (The explanation (*aition*) is that it is their *nature* (*pephuken*) to go there . . .) is not the assertion of a cause in the strong sense of an agent or agency, but should at least, consistent with *Phys.* II, put an end to the causal chain. To ascribe something to a thing's nature is to give a self-sufficient explanation, to make the explanation the equivalent of an irreducible law of nature in modern physics. Why did the water go up? Because it was changed into air. Why did the air go up? Because it was light. Why does the light go up? Because that is its nature; i.e. that is the ultimate behaviour of light things. If one asks further why that is the behaviour of light things, it will show only that one does not understand what Aristotle means by saying that is the nature of the light. Similarly, if one asked Newton why bodies in motion tend to stay in motion and bodies at rest tend to stay at rest, one would show only that one did not understand his assertion that this is a law of nature; if one pressed him for more explanation, he would reply, 'Hypotheses non fingo.'

The problem with Aristotle's account here is not his explanation at ^{b}15 but the fact that he persists in seeking an answer to what he should say is the unanswerable question of what air is moved *by*, and in assuming that the answer provides important information about the motion of air. Note that if Aristotle were right in his present analysis of motion, he would be forced to abandon his distinction between nature and power. Nature is an internal source of change, power an external source of change. But it will turn out that every nature is really a passive power; hence there is really no internal source of change in the world, only patently external sources of change and latently external sources of change. Some external agent is needed to originate motion even in natural substances. Thus a basic Aristotelian distinction between natural and non-natural motion collapses. This point is perhaps not decisive against Aristotle's present analysis, but it does show how radical his proposal is relative to his own stated principles of change.

255^{b}31–256^{a}3

Aristotle is now in a position to make the general assertions which will produce the desired conclusion:

ARGUMENT C

(1) All things in motion are moved either by nature or by force.
(2) All things moved by force are moved by something.
(3) All things moved by nature either move themselves or they do not.
(4) Things that move themselves are moved by something.
(5) Things that do not move themselves are moved by some-

thing, either by what creates them or by what removes the obstacles to their motion.

(6) Thus, all things in motion are moved by something.

Suddenly Aristotle has switched his argument from that of the first paragraph, argument A, to something like the alternative argument sketched above, argument B. But argument C differs from argument B in that Aristotle has dropped his appeal to the distinction between objects moved by themselves and those moved by another—a distinction which only interfered with his argument. As we anticipated, most of the chapter has been devoted to point (C5), the analysis of simple natural motion. We have seen strong reasons for rejecting (C5)—reasons growing out of fundamental Aristotelian principles. Nevertheless, Aristotle seems willing to slight some of his own distinctions in the theories of nature and cause in order to arrive at his conclusion, which will ultimately lay the groundwork for a new understanding of the causes of motion.

After giving a sympathetic interpretation of Aristotle's argument, Simplicius concludes his treatment of the present chapter with the following remarks:

One might be puzzled, I think, about how Aristotle can say that bodies that are in natural motion in place are moved by an agent that moves those very bodies, since the creator and maker of the substance both makes the form and is the cause of the generation, although generation is different from change in place. And in general, if what creates and makes fire ceases acting at some time and neither is present nor in contact with the fire when it is in motion, how can fire be said to be moved by it in such a way that he can conclude that every body in motion is moved by something other than the moving body?

This must be investigated further, but let what has just been said suffice for now: what can create the form of the naturally movable body is in some sense able to move it with its natural motion. For it makes both the motion and the actuality of the form along with the form, as what makes the amphora also renders it able to contain the wine. Nevertheless, what makes the motion along with the form does not in the proper sense impart that motion to the form, as e.g. does one who imparts motion to a stone with his hand, since the maker is no longer present when the body is set in motion; but this appears as a further specification of a mover, according to which the mover makes a thing to be naturally capable of moving and to be in motion if nothing hinders. And amazingly (*thaumastōs*) Aristotle discovers even in things undergoing natural motion that what is moved is moved by something, since to be in motion is to suffer some change (*paschein ti*), and what undergoes change requires an agent. For the perfect actuality proceeding from the perfect substance does not require any other productive cause, but motion, being an imperfect actuality that is rather passive and mixed with much that is potential, needs something to produce it. (*In phys.* 1220. 5–26)

Simplicius does a much better job laying out the *aporia* than in solving it—even in a preliminary way. To make the element is in some sense to confer on it motion; but it is not to *impart* motion. But is what is conferred something more than just the potential to move, or, more properly, the nature—the inner source of motion? If not, we cannot say that the element is moved by something in a univocal sense of 'move'. If so, what is the sense? I do not see that Simplicius ever fulfils his promise to reconsider the question.

Modern scholarship has not done much better. Recent attempts to account for the natural motion of elements proceed by denying that nature includes a principle of rest (Gill 1989: 236–40, esp. 238 n. 60) or, alternatively, a principle of rest (Cohen 1994). Bodnár (1997) effectively criticizes these attempts (90 ff.), but in their place offers more of a description of Aristotle's views than a justification. Indeed, Bodnár gives an especially compelling account of the problems created by Aristotle's view, most notably the point that any theory that depends on causes which generate the element or remove obstacles to its motion will have serious trouble accounting for the motion of the fifth element, which can have neither kind of cause (109 ff.). Furthermore, it remains unclear how the unmoved mover can fill the explanatory gap (pp. 116–17).

Concluding Remarks

In general, one is left with the problem that the only way to save Aristotle's concept of natural motion is to take seriously his remarks which tend to make the contribution of the external mover an incidental one. On the other hand, to make the contribution of the external mover incidental is to subvert the claim that every body in motion is moved by some agent—where that agent is understood to be, as Aristotle seems to wish, an agent distinct from the body in motion. For if bodies describing natural motions are moved only incidentally by a distinct agent, they are not intrinsically moved by that agent. Hence the principle he wants to establish here, which will be essential to his argument for a first unmoved mover, is not established. To establish his principle that every body is moved by a distinct agent, he will have to make the contribution of the agent that creates the natural body, or the one which removes an obstacle to its action, intrinsic. This would be to ascribe to the agent an active power and to the natural body a corresponding passive power, a possibility which Aristotle also seems to adumbrate in the chapter. But if that is so, nature is not really distinct from power, but is rather a certain kind of passive power; and it is no longer clear how natural motion is self-motion.

CHAPTER 5

At the end of the last chapter Aristotle established that every body in motion is moved by something. But he left the 'something' which is the cause of motion unclear. In the present chapter he will argue for three important theses about the something: (1) however many movers there are in a series of causes, there will always be a first mover; (2) the first mover will in some sense move itself, and hence be a self-mover; (3) self-movers consist of two essential parts, an unmoved and a moved part. Aristotle set himself on a path towards a metaphysical analysis of physics in Ch. 4, though modestly, with the apparently innocuous claim that everything moved is moved by something. Here he will be making ever bolder analyses.

256ᵃ5

'that moves the mover' (*ho kinei to kinoun*): the Greek phrase could mean: 'that the mover moves'. However, the ensuing discussion clarifies the distinction. Aristotle makes two distinctions here: that between (a) movers that do not initiate motions and (b) movers that do, and a subdivision of (b) into (i) immediate movers and (ii) remote movers. In the present paragraph Aristotle argues that (a) presupposes (b), and in the next paragraph he will argue that (ii) presupposes (i). The priority of (b i) will bring us close to the priority of the self-mover.

256ᵃ10

'For that moves the last, but the last does not move the first': one might object to Aristotle that it is no less true that the first mover needs intermediate movers than that intermediate movers need a first mover. If Aristotle should reply that e.g. the man in his example could move the stone with something other than the stick, the objector could reply that something other than the man could move the stone. But, Aristotle could counter, the man does not need an intermediate mover at all: he could cause motion by his hand alone. To this the objector might reply that in such a case the man would be using his hand as a mere instrument; like the physician who heals himself *qua* other, the thrower uses his throwing arm *qua* other. Shortly Aristotle himself will make distinctions between mover and moved in the self-mover; it is not clear how he can block such a move by an objector. Aristotle's fundamental outlook here seems to hark back to Plato's analysis of cause in the *Phaedo* (esp. 99a–b), according to which the real causes are things endowed with intelligence (*nous*). Physical bodies are at best concomitant causes (*sunaitia: Tim.* 46c–d), fit only to be instruments in the service of some higher, teleological agency (cf. *Tim.* 46e, 48a). But he does not seem have made his point thus far.

256ᵃ17

If there is no first mover, there will be an infinite regress. (Ross takes this remark as a parenthetical one explicating only why (b) does not presuppose (a). But the explanation also gives the only real justification for the previous point that (a) presupposes (b)—ᵃ10–13 advances a weaker claim.) Aristotle states his point badly: that there is no first element does not explain why it is impossible to have an infinite series, but only gives a condition for there being an infinite series. Aristotle's main point is that if we do not require closure to the series of causes, we shall generate (at least the possibility of) an infinite series; but there can be no actual infinite series; hence there must be a first cause, a first mover. Aristotle could simply appeal to arguments elsewhere to rule out an infinite series. In the present context, however, we should not let the rejection of infinite series pass unchallenged. One key argument against the infinite series is that it is impossible to cross an infinite series in a finite time. But Aristotle holds that the world is everlasting, and that time has no beginning. Why, then, should we not allow the possibility of an infinite series of efficient causes in the sublunary world which recede back through infinite time? Indeed, how could the series of efficient causes in the sublunary world *not* be infinite?

It is true that between any point in a causal chain and the outcome of that chain there will be a finite number of steps. But why should there not be an infinite number of steps preceding the outcome? Indeed, if Aristotle admits events such as the creation of air from water, as he seemed inclined to do in the last chapter, as part of the causal story, why should we not be able to trace the causal chain back through previous transformations of the elements throughout the ages? Aristotle may be able to maintain that there is a finite number of 'vertical' stages to the cause of elementary transformations: namely, the sun in its annual course (*GC* II. 10); but it is difficult to see how he can posit a merely finite number of 'horizontal' stages among efficient causes of the same level. (A similar problem arises for Aquinas's second proof for the existence of God, through efficient causes: *Summa Theologica*, pt. I, q. 2, art. 2.)

256ª19–21

Having now established that there is a first mover in every causal sequence—a thesis which he does not highlight at all—he quickly moves on to his more explicit objective: the claim that the first mover moves itself. The compressed argument assumes that the first mover is moved either by an agent external to itself or by an agent not external; by elimination, the first mover is moved by an agent not external—i.e. by itself.

Let us reconstruct the argument of the present paragraph.

(1) Everything that is in motion is moved by something (from Ch. 4).

(2) Everything that is moved by something is moved either (a) by something that is moved by another or (b) by something that is not moved by another.

Now at this point we need a clarification. In (2), are we talking about the immediate or proximate mover or about the ultimate mover? Initially, Aristotle seems to focus on the former, as he discusses the implications of a chain of movers. But because the chain cannot consist solely of movers of the type referred to in (a), we are led to recognize that a mover of the type referred to in (b) is necessary for every chain of causes, i.e. as an ultimate mover.

(3) If all movers in the series are moved by another, there is an infinite regress.

(4) There is no infinite regress.

(5) Thus, not all movers in the series are moved by another.

Whenever we reach a mover of type (b) the series of causes ends, so that mover becomes, in effect, a first mover. By definition, then,

(6) The first mover is not moved by another.

As pointed out in the previous section, Aristotle does not seem to have a good reason for advancing (4) in the present case. But leaving that problem aside, we see that Aristotle can advance from causes of type (a) above to causes of type (b).

In his argument (last sentence of the paragraph) Aristotle advances to a further consideration of the first mover.

(7) Suppose the first mover is moved.

(8) Thus, the first mover is moved by something (2).

At this point we need a premiss that looks superficially like (2), but which has different disjuncts:

(9) Everything that is moved by something is moved either (a) by another or (b) by itself.

(10) Thus, the first mover is moved either (a) by another or (b) by itself (9, 8).

(11) Thus, the first mover is moved by itself (10, 6).

We finally arrive at Aristotle's conclusion, but only by introducing statement (9), which says something rather different from (2). Whereas (2) seems to address immediate movers which themselves have, or may have, movers in a chain of causal antecedents, (9) presents only the alternatives for any immediate cause of motion. Furthermore, in (2b), Aristotle refers only to what is not moved by another, giving no positive characterization of that alternative. Although Aristotle may think of (9) as somehow implicit in (2), it does not follow logically from the earlier premiss, so needs to be introduced separately. If pressed to defend it, Aristotle could claim that it was self-evident, presenting as alternatives the only logically possible cases.

Notice that the conclusion can only be a conditional one, based as it is on supposition (7):

(12) Thus, if the first mover is moved, it is moved by itself (7–11).

In the last sentence of the paragraph Aristotle duly presents his conclusion as a conditional one, citing both our (1) and (7) as conditions—though he has already attempted to prove (1) in the previous chapter. By making (7) a condition, Aristotle leaves open the possibility that the first mover is *not* moved. Ultimately he will argue that all moved movers presuppose an unmoved mover, so the present proviso is an important one. For now, Aristotle can claim to have proved that every chain of causes goes back to a first mover, which must be either self-moved or unmoved.

256ᵃ21–ᵇ3

The previous argument is built on the construction 'X is moved by Y', where Y denotes an agent. The present argument focuses on the construction 'Y moves X by means of Z', where Z denotes an intermediate mover. The argument is analogous: though there can be a series of intermediates, the series cannot go on to infinity; hence there must be a mover that moves by means of itself. Schematically, every motion in which x is moved by y by means of z, ($Mxyz$), presupposes a case in which there is some w such that w is moved by w by means of w, ($Mwww$).

256ᵇ3–12

Aristotle adds another argument: if every mover is moved incidentally, then it is possible that at some time nothing moves. But it is not possible that at sometime nothing moves; hence not every mover is moved incidentally—some mover is moved intrinsically. This argument is reminiscent of the argument at the beginning of *Met.* Λ6, where Aristotle argues that since substances are the ultimate realities, if all substances are destructible, then all things must be destructible. But if all things are destructible, it is possible that at some time nothing existed. Moreover, given that in an infinite period of time all possibilities are realized, the possibility that nothing exists will be realized (cf. *Cael.* I. 12, 281ᵇ21–2). But if at some time in the past nothing existed, nothing exists now. Yet it is false that nothing exists now; hence not all substances are destructible. In the present argument, what is in question is not the destructibility of substance but the possibility that all motions are incidental. The form of the argument, however, is similar: from the hypothesized possibility we deduce a possibility that there is no motion. For on this account there is no necessity that anything move. But that possibility is incompatible with a demonstrated impossibility; hence the hypothesis is false.

Some ancient commentators known to Alexander wished to locate this paragraph after 258ᵃ8 (Simpl. 1224. 6 ff.). But the passage prepares for 256ᵇ27 ff. and can only precede it.

256ᵇ12

'For it has been proved earlier': in Ch. 1.

256ᵇ13–27

Aristotle adds a further argument, one intended not as a demonstrative proof, but only as a persuasive analogy. Dividing the bodies in a causal sequence of motion into mover, means, and moved, we can associate the second with causing motion and being moved, the third with not causing motion but being moved; but with what shall we associate the first? With causing motion but not being moved. The conclusion of this analogy turns out to be quite different from the previous demonstrative argument: here we conclude that the first mover is *not* in motion; previously we concluded that the first mover *is* in motion—intrinsic motion, but motion none the less.

257ᵃ3

'Or it may not happen like this': the previous argument that the mover cannot be moved with precisely the same movement as the moved is straightforward. The second argument is less perspicuous. Why should the only alternative to the same movement be a succession of different kinds of movement? For instance, according to Newton's second law, to every action there corresponds an equal and opposite reaction. Thus the recoil of a cannon has an equal force to the cannon-ball, in an opposite direction. Here the 'mover' has the same *kind* of motion as the moved, but in a different direction—in Aristotelian terms the action is numerically different. Why should we rule this out a priori? It seems to be empirically true and in accordance with the laws of conservation of motion and energy that physical bodies move either with the same kind of motions which they impart to other bodies (e.g. in the collision of elastic bodies) or with different kinds of motions preserving the same energy (e.g. collision of inelastic bodies generating heat).

Aristotle assumes that the only real alternative is a succession of kinds of motion, of which there is a finite number (ᵇ7). Strictly speaking, there are only three genera of motion according to Aristotle: namely, change in quality, change in size, and change of place (*Phys.* V. 1). But in his illustrations, Aristotle introduces a number of what we might call 'species' of motion: e.g. healing, teaching, throwing. Are there a finite number of these species? It is not clear. For instance, are overhand and underhand throwing subspecies of throwing? And are fast-ball throwing and curve-ball throwing sub-subspecies of overhand throwing? If the criteria for subspecies are indeterminate, it seems possible to generate an infinite number of subspecies. But assume for the sake of argument that there is a finite number of changes. Aristotle examines the two possibilities. A series of different types of change will come to an end; hence we shall need a different account of the first mover. The other possibility is a 'return to the beginning of the series': i.e. a repeating cycle of types of motion. Let there be three genera of motion. If *a* moves *b* moves *c* moves *d*, where *a* has motion type *X*, *b* type *Y*, *c* type *Z*, then, according to Aristotle, since there is a succession of types of motion, *d* will again move something with motion of type *X*. But since any member of the series *a–c* is a mover of *d*, and the earlier mover is more of a mover than the later, *a* will move *d*; moreover, *a* will move *d* with motion *X*. As Ross points out, the reasoning is fallacious. Although *a* is a mover of *d*, *a* does not move *d* with motion *X*, since *a* does not immediately move *d*. For example, if Al

pushes Bill, and Bill trips Carol, and Carol hits Dale, it may be true
in some sense that Al moved Dale, but it does not follow that Al
pushed Dale. Hence the case of alternating motions does not reduce to the case of the mover moving with the same motion as the
moved.

257ᵃ22

'as we said previously': see ᵃ7–12.

257ᵃ27

Having now established that the first cause of a series is in some
sense self-moved, Aristotle turns to an analysis of how self-motion
is possible, the new enquiry constituting the second major division
of his argument in Ch. 5.

257ᵃ31–3

Having dismissed the possibility that the original mover is moved
by another, Aristotle focuses on the possibility of its self-motion.
He will argue that (A) the original mover does not move itself as a
whole (ᵃ33 ff.), and (B) that different parts of it do not reciprocally
move each other (ᵇ13 ff.); (C) neither do proper parts move themselves (ᵇ26 ff.); but (D) it will be composed of a part that is an
unmoved mover and a part that is moved (258ᵃ5 ff.).

257ᵃ33–ᵇ13

(A) By invoking distinctions from *Phys.* VI, Aristotle can argue
against the original mover moving itself as a whole: if it did, it
would as a whole be both causing and suffering the same change.
We have seen above (256ᵇ34–257ᵃ3) that the same subject cannot
simultaneously be causing and suffering the same action. At
257ᵇ6 ff. Aristotle further strengthens the argument by invoking the
actuality–potentiality distinction and his definition of motion in
Phys. III. 1 to show that the subject of the alleged self-change

would simultaneously have to both have the character F which it is causing and not have it, in so far as it is being caused to be F.

257^{a}33–^{b}1:   Cf. *Phys.* VI. 4, 234^{b}10–20. The infinite divisibility of the movable seems irrelevant here, since the kind of parts which Aristotle will ultimately invoke in his analysis of self-motion are not spatial ones.

257^{b}6:   'it is in motion through potentiality ...': translation after Waterlow (1982: 244). As she argues (n. 27) the obvious (and usual) translation does not work here. For example, Hardie and Gaye's 'this is potentially, not actually, in motion' gets the point wrong. The body in question is *actually* in motion, but its actuality is that of a potential mover, or, as Aristotle puts it, an incomplete actuality (^{b}8); otherwise the actuality would be the completion of the process, not the process itself. For instance, the actuality of the buildable, i.e. of that which has the capacity (potentiality) to be built, is the building process, while the actuality of that process is, say, a house. See *Phys.* III. 1 and Graham (1989).

257^{b}11:   'and likewise with everything ... same name as the moved': on the 'Synonymy Principle of Causation' see Barnes (1979b: i. 119), who finds the principle in Xenophanes (p. 88) and Alcmaeon, as well as in Aristotle and Descartes (*Meditation* III). According to the principle, a cause can impart only what it itself has. In Aristotle's pluralistic account of causation not every kind of cause is synonymous with its effect: Theophrastus (from *Phys.* III *ap.* Simpl. 1236. 1 ff.) cites e.g. the Sun as a cause of generation (man produces man synonymously, but the Sun produces man non-synonymously; cf. *Met.* Λ5, 1071^{a}13 ff., ^{a}20 ff.) and a whip as the cause of welts. If synonymous causation does not exhaust the field of causation, the present argument cannot suffice to show that part of the self-mover causes motion and part is moved, for all self-movers. For there may be self-movers which are not synonymous causes. Aristotle may mean to distinguish between mover and moved only for synonymous causes in ^{b}12–13. But his conclusion still seems too hasty, for, by invoking the law of non-contradiction (see *Met.* Γ3, 1005^{b}19–20) as his grounds for rejecting self-motion of the whole, he leaves open the possibility that the subject causes motion and is moved in different respects rather than in different parts. In any case, Aristotle is just beginning his analysis of self-motion.

257^{b}13–26

(B) Different parts of the original mover do not move each other.

ARGUMENT B1 (^{b}15–20)

(1) If they did, there would be no first mover. For example, if A moves B and B moves A, we cannot identify either one as prior to the other.
(2) Hence there is no first mover.
(3) But it has been proved that there must be a first mover.
(4) Hence this option is impossible.

This argument is unpersuasive. Even if there were no priority in the *parts* of the original mover, it would still, as a whole, be the first mover. It is not yet clear what sort of parts we are dealing with and, consequently, what sort of properties they might instantiate. Aristotle might also wish to complain that the present alternative involves a vicious circle within the original mover, but he does not raise this problem; nor is it clear that the reciprocal causation posited is viciously circular.

ARGUMENT B2 (^{b}20–3)

(1) The original mover need not be moved except by itself.
(2) Hence, if the other part sets it in motion, it does so incidentally.
(3) Hence, the other part does not contribute to the mover's causing motion.
(4) Hence, the first part is unmoved, the second moved.

There seems to be a serious ambiguity in 'original mover' in the first statement, or at least a serious confusion as to what the 'other part' of statement 2 is other than. If we take the 'original mover' to be a privileged part of the self-mover, then we have already begged the question about what the mover is. Surely (1) could be true where 'the original mover' means 'the whole self-mover'; yet we could leave open the question whether proper parts A and B of the whole set each other in motion. Suppose A sets B in motion and B sets A in motion, according to the hypothesis that Aristotle is examining here. Why should that violate (1)? A and B are parts of the whole mover; hence for one part to set another in motion is not for something outside the whole to move it. An advocate of the present view would not want to claim that A itself was a self-mover, and hence would have no reason to say that A could not be moved. Statement (3) clearly follows from statement (2). But (4) does not follow from anything said so far: Aristotle would certainly need some further premisses concerning the non-A part of the whole to show that some part has to be moved.

ARGUMENT B3 (^{b}23–5)

It is difficult to know what the point of this is. The appeal to everlasting motion is too general: what about it rules out A's being moved? Surely we cannot appeal to the cosmological theorems of the later part of this book yet; they have not been proved. Cornford and Wicksteed translate as if the argument were a reply to an (obscure) objection to Aristotle's own account. But why, then, should Aristotle put it in a series of positive arguments against an opposing view? Simplicius interprets the argument as saying that the everlasting motion of the cosmos (proved in Ch. 1) would not be possible on the present hypothesis, because it would imply an infinite regress. On his view, the present argument would have much the same point as argument (B1). Wagner takes it as a rejection of a 'theorem' according to which every process must have a contrary process.

ARGUMENT B4 (^{b}25–6)

Aristotle's appeal to argument (A) also needs clarification. If X causes Y to be in motion M and Y causes X to cease M (be in not-M—only one of several possible modes of their interaction), Aristotle might complain that the original mover composed of X and Y is exemplifying both M and not-M. However, Plato had already made it clear (*Rep.* IV, 436b–e) that alleged contradictions can be dissolved by judiciously assigning contrary properties to different parts of a thing; there is surely then no need to attribute conflicting properties of different parts to the whole as a unit. Taking the internal combustion engine as a self-mover, the fuel supplies heat energy, which the pistons convert to vertical motion, which the crankshaft converts to circular motion, while some circular motion is converted elsewhere to electricity, which supplies the spark to convert more fuel into heat energy. Each part has its

own work to do without thereby acquiring contradictory properties. In the absence of more details, the present argument does not
seem to constitute a compelling objection to reciprocal causation of
parts.

257^{b}26–258^{a}2

(C) The original mover is not moved by virtue of either (1) a part
moving itself or (2) the whole moving itself. If (1) a part moves
itself, then that is the primary mover, not the whole. This seems to
be a sound argument. Argument (A) seems to have already ruled
out (2), but here Aristotle stresses the fact that on the present
analysis the parts need not be in motion at all. The point is well
taken, but how can Aristotle infer from this point that some part
must be unmoved, another moved? (Simplicius takes the conclusion to result from eliminating all alternatives, i.e. as a global
conclusion of the whole section; but in the text the statement reads
like a conclusion only to the present argument.) The obvious conclusion is, rather, that the whole may move without the parts moving. In one sense this is quite true: in a rotating sphere the parts
rotate only incidentally (they are at rest relative to each other and
the whole), whereas the whole sphere is in motion. Aristotle does
succeed in showing that if the whole is moved by the whole, the
alleged self-motion of the parts cannot explain the overall self-
motion. But he does not really succeed in showing that the whole
must consist of an unmoved and a moved part, because he does not
succeed in eliminating all alternatives.

258^{a}3–5

'Further, if the whole moves itself . . . and by A alone': these two
sentences have caused problems since antiquity. Do they conclude
the previous argument (C), or do they introduce a new argument?
The passage makes more sense as an introduction to the following
argument; but the argument is not an afterthought: it provides a
positive argument (D) that the self-mover is composed of an unmoved and a moved part. So far, the conclusion has been arrived at
only through an argument by elimination, and a rather unsatisfac-

tory one at that, as we have seen. The positive argument supplies important confirmation.

258ᵃ5

Aristotle produces a taxonomy of four classes by dividing (a) movers into (i) moved and (ii) unmoved and (b) moveds into (i) movers and (ii) non-movers. We are apparently expected to conflate class (a i) with class (b i) to come up with a threefold division of possibilities which will be represented by A (= a ii), B (= a i/b i), and C (= b ii) in the following argument.

258ᵃ9–18

C is not necessary to the self-motion of ABC, because we can detach C and still have motion: namely, that exhibited by B. On the other hand, A is necessary as moving cause, and B as first moved object. Although Aristotle says nothing about the special relationship between A and B, clearly there must be one, because somehow A can move B even without itself being in motion. Hence there must be some kind of non-physical causation between them. Since B is moved, it can interact with C in a physical way that requires no special explanation. The relation between A and B remains mysterious.

Aristotle's conception—and the problems inherent in it—may be illustrated by Cornford's example, where A = soul, B = body, C = clothes. The conditions are satisfied: AB is a self-mover without C. But ABC is not a real unity to begin with. Take a more reasonable candidate for a natural unity: A = soul, B = heart, C = body. Now it is true that for the body to be moved not incidentally, the heart will have to move it (according to Aristotelian theory); but what would it mean to say that we can leave C out of the whole? Is AB a real unity of which we can say it moves itself? The heart is not viable apart from the body, so the soul–heart conjunction is not an organism of which it seems appropriate to say that it is a self-mover, as if it had some sort of independent existence. We must hold, then, that B is not a proper candidate for being the first moved part. Perhaps only the body will qualify for that role. The example shows that we must be careful about how we define the mover and the first moved. It remains to be seen if any definition could be successful.

258ᵃ20–1

It is possible for the mover to 'touch' the moved while the latter does not touch the former: *GC* I. 6, 323ᵃ25–32; presumably this is true not merely in virtue of linguistic usage (the only reason which Aristotle gives there is that we say someone who grieves us 'touches' us—a pun—ᵃ32–3), but because in such cases the mover does not have matter: cf. ibid. 7, 324ᵇ3–6, 10–13. Simplicius (1243. 25–8) remarks that an immaterial unmoved mover can touch the moved part not in the proper sense, but only metaphorically. Perhaps we can best take 'touching' as a term for causal action, with one-way touching as causal action in a single direction, mutual touching as mutual interaction.

258ᵃ22

'each part will touch the other': this phrase is found in only one manuscript, and is omitted by most editors and translators. But, as Ross notes, it alone makes the conditional clause relevant.

258ᵃ27–ᵇ4

If a self-mover is continuous and we remove part of it, what will happen? If the remainder is no longer a self-mover, there is no problem. But if it is a self-mover, then it appears that only the minimal part that remains is the real self-mover. This aporia is a typical thought-experiment to challenge Aristotle's theory. There seems to be a simple way out: to declare it an empirical question just what constitutes a continuous self-mover. If we find that by removing some portion of it, it still moves itself, we limit the identification of the self-mover to that portion. We repeat the experiment until we find some minimal portion than which there is no smaller portion which is a self-mover. The procedure is in accordance with the method of the *Posterior Analytics* (introduced in I. 4) of identifying a subject that is commensurately universal with a given attribute. This procedure would be problematic, how-ever, if we were dealing with a completely continuous substance such as water, which will have its own natural motion no matter how small the sample. Aristotle has already ruled out such sub-stances as self-movers in Ch. 4, in part because of their very simplic-ity, so perhaps we are expected to envision a more complex kind of self-mover.

Aristotle's actual solution does support this reading. If the al-leged self-mover, by being divided, loses its nature, then it was a real self-mover. For instance, if I cut a lizard in half, I have in a sense divided it, but I do not preserve its viability. The problem with this example is that we do not really fulfil the hypothesis that it still be self-moving after the division. Suppose, instead, I break off the lizard's tail. The lizard is still a self-mover (the tail of course is not). Does that mean that the lizard minus the tail is the real self-mover? Here one could say that what we have is an incomplete lizard, a deformed lizard, i.e. a less than perfect specimen. It seems false in this case to say that the lizard has lost its nature, though it is true to say that it has lost part of its capacity. The parallel is perhaps close enough to say that it supports Aristotle's general claim here by showing that there are circumstances under which a self-mover could still be considered a proper self-mover even if some proper part of it could move itself. A more difficult case would be that of a flatworm which, being cut in half, produces two flatworms. One could say that the original flatworm lost its nature of being a single flatworm; but one could not say that it lost its nature in the sense of the species form it exemplifies.

All of this must remain mere speculation on what Aristotle has in mind here, for the next chapter will undermine the claims of ani-mals to be self-movers. At this point Aristotle is engaging in dialec-tical manœuvres which will allow him ultimately to advance from self-movers to the first unmoved mover as the focus of his study.

258^{b}4

Having started from the conclusion of Ch. 4 that every moved object is moved by something, Aristotle has now arrived at his conclusion, the thesis of Ch. 5, that the original mover is unmoved.

For even if the original mover is a self-mover, it is divisible into a moved part and an unmoved part. The present thesis is the premiss on which the remaining argument of the book will depend.

Concluding Aporia

Aristotle has used numerous distinctions in Chs. 4–5 to further his analysis of the causes of motion and of self-motion. He employed the intrinsic–incidental distinction in both Chs. 4 and 5; he used the potentiality–actuality distinction and the different levels of actuality in Ch. 5; he distinguished natural and forced motion in Ch. 4, as well as self-motion and motion caused by another; in Ch. 5 he has distinguished the roles of the whole and the part in self-motion. One distinction conspicuous by its absence from the present discussion is that of the four causes. Ultimately, that distinction will be very important for understanding how something unmoved can cause motion. Why does Aristotle not introduce the distinction? One possibility is that so far only efficient causation has been under consideration. Certainly the examples Aristotle adduced in Ch. 4 to elucidate elemental motion support this conjecture: the agent that makes water into air as well as the agent that removes a lid are paradigmatically efficient causes. Furthermore, the 'moved by' locution prominent in Chs. 4–5 at least immediately suggests the activity of an efficient cause: we say that the horse is moved 'by the rider', or 'by its desire for food', rather than 'by health' (final cause), 'by equinity' (formal cause), or 'by its body' (material cause).

One problem that arises is whether an argument built on an analysis of efficient causes will be able to arrive at a first mover that is something other than an efficient cause. If X is moved by Y and Y is moved by Z, and Z is the original cause and also not an efficient cause, will the 'moved by Z' locution be equivocal? Clearly, the four causes represent four 'becauses', different senses in which one thing is responsible for another (see on 1, 252^b4). One cannot, in general, substitute one kind of cause for another in an explanation and preserve the truth-value of the account. Suppose that Socrates is in prison because he thinks it best to obey the decision of the court (final cause) but in another sense because his bones and sinews are so arranged (material cause); the second description is compatible with Socrates' taking flight tonight, but the first is not. There is, then, at least a potential problem with introducing one sense of causation into a context employing a different sense, like switching horses in mid stream. Can Aristotle make good on his argument that every motion presupposes an unmoved mover?

CHAPTER 6

258^b16

Aristotle in fact seems to allow some types of things to appear and disappear without coming to be and perishing: namely, things which are simple. These include sensations (*De Sensu* 6, 446^b2–4), certain causes and principles (*Met.* E3, 1027^a29–30), mathematical points ('if they exist', *Met.* H5, 1044^b21–2), and forms and essences (*Met.* Z15, 1039^b20–6; H3, 1043^b14–18; 5, 1044^b21–3). Coming to be and perishing are processes which require stepwise changes of subjects having parts; simple subjects cannot enter into such changes.

258^b20–6

The argument is difficult to follow. Ostensibly, it seems to go thus:

(1) Suppose some unmoved movers alternate between existence and non-existence without coming to be or perishing.
(2) Suppose these unmoved movers are self-movers.
(3) Nothing without parts can be in motion.
(4) Thus, everything that is moved has some size.
(5) It is not necessary that an unmoved mover have some size.
(6) Thus, there is some cause of (1).
(7) Thus, there must be some unmoved mover that does not alternate between existence and non-existence.

First note that 'unmoved' in the present argument does not mean absolutely unmoved, but merely not moved by some prior mover. What Aristotle has in mind are self-movers rather than absolutely unmoved movers. Statement (3) makes sense within Aristotelian theory: size is a prerequisite for moved bodies (see *Phys.* VI. 4), and from this (4) follows. Size is not a prerequisite for unmoved movers, as (5) correctly asserts. But how do these points entail (6)? We might wish to infer the nature of the cause referred to in (6) from some facts about the events caused. But how do (4) and (5) further that enquiry? We might show how the property of size exhibited by self-movers disqualifies them from being ultimate causes. But in fact we already proved in the last chapter that there is some unmoved component to a self-mover. And (5) in fact serves to dissociate the unmoved component from the material extension of the moved part. Thus, in effect (5) makes us look for a different reason for disqualifying the unmoved component as a cause of the alternation between existence and non-existence of self-movers. One obvious reason why the unmoved component cannot be a cause of its existence or non-existence is simply that it participates in the change itself. But then why bring in (4) and (5) at all? Note also that whereas we had to construe 'unmoved mover' in a broad sense to include self-movers in the first two premisses, we must construe the phrase strictly in (5). Is there an equivocation going on?

Or perhaps the conflict between the hypothesis and (5) will be just what we need to make Aristotle's point: in order for an unmoved mover to be and not be without a process of coming to be, it must be metaphysically simple. And, to cause its own change, it must be a self-mover. But to be a self-mover, it cannot be simple for the reasons given in (3) and (4). Thus it cannot cause its own change, and we must posit some prior cause. This line of reasoning will get us to (6), but not via (5). We need something like:

(5a) It is not possible that an unmoved mover satisfying (1) have some size.

And, we should add:

(5b) The (allegedly) unmoved mover will in fact undergo a process of coming to be which of course will require a cause.

By such a route we can get to (6), though it is not, alas, clear that this is what Aristotle has in mind. Simplicius takes the argument in a way similar to the one we have just sketched: he understands Aristotle to use (4) to demonstrate that the self-mover will, after all, undergo coming to be and perishing, since as a whole it has parts.

We might look for further enlightenment in the following discussion (^{b}26 ff.), but there Aristotle shifts (almost imperceptibly) from talk of the coming to be and perishing of self-movers to the coming to be and perishing of things in general. Thus the following discussion seems to constitute a different argument.

258^b25-6: 'there is no need for the mover to have magnitude': Aristotle will eventually prove that the first unmoved mover is without magnitude (10, 267^b18-19).

$258^b26-259^a6$

This argument seems to proceed as follows:

(1) There is continuous and everlasting coming to be and perishing in the cosmos. (Let this be called 'cosmic change'.)
(2) There is some cause of cosmic change.
(3) Perishable things, whether taken individually or collectively, are not everlasting or necessary.
(4) Cosmic change is everlasting and necessary.
(5) Thus, perishable things, whether taken individually or collectively, are not the cause of cosmic change.
(6) Thus, there is some prior cause of cosmic change.

There is some parallel with the argument at $^b20-6$, but clearly the initial premiss here deals with coming to be and perishing in the cosmos as a whole, rather than in the domain of self-movers. For point (4) Aristotle can draw on his Principle of Plenitude, according to which what is everlasting is necessary (see on 1, 251^a17). Point (3) slides too easily from the individual to the collection: although it is a tautology that individual perishable things are not everlasting or, consequently (on Aristotle's theory), necessary, why could not the collection of perishable things be everlasting? In fact, Aristotle holds that species of animals and plants are everlasting (e.g. *GA* II. 1, $731^b24-732^a1$), a clear instance of the fact that the class may have properties which the individual lacks.

There is a further problem, in that an additional principle is needed to make the argument work. In order for (3) and (4) to lead to (5), we must have a general proposition to the effect that the properties conferred on the effects must be found in the cause. This brings us back to the Synonymy Principle of Causation (see on 5, 257^b11 above). But, as we have noted, not every cause confers its own character on the effect. Aristotle needs to tell us what sorts of properties are preserved, or under what circumstances they are preserved across a causal relation. It is not clear what the solution would be to the problem, but we may be fairly certain that Aristotle would hold that everlasting existence and metaphysical necessity are properties that can be imparted to phenomena only by beings that already possess them.

Unfortunately, having established that perishable things cannot cause cosmic change, Aristotle does not specify what kind of cause would be responsible for such change. Presumably we are expected at this point to recall the arguments of the last two chapters, to tell us that it will belong to the class of unmoved movers.

259ᵃ9–12

Ross refers to his note to *Phys.* 188ᵃ17–18, where he points out the scientific advantage of explaining with the fewest possible principles, an anticipation of Ockham's razor. True, but if we focus exclusively on this point, we will miss the vast difference between Aristotle and modern science. For Aristotle, following Plato (*Phd.* 97c–d), the world is organized for the best (cf. here *to beltion*, ᵃ11). For Aristotle the point is supported not by a concern for the most austere methodology but by a commitment to the most optimistic value theory. For Aristotle the good is associated with order, and as social-political order is personified in a leader (e.g. the general of an army), so natural order is realized in a first principle (*Met.* Λ10, esp. 1075ᵃ11–19). That is, nature is arranged like the best social system we can imagine. The ideal is archaic, as shown by Aristotle's quotation of Homer, *Iliad* 2. 204, in the last line of *Met.* Λ10: 'The rule of many is not good: one ruler let there be.'

259ᵃ12–13

Nothing Aristotle says here is incompatible with the plurality of unmoved movers in *Met.* Λ8. There he retains a first unmoved mover which is prior to the many movers of the spheres. On the other hand, the present passage is compatible with there being no plurality of movers. Aristotle's preference is clearly for a single unmoved mover. This passage can thus be taken as evidence that *Phys.* VIII was composed before *Met.* Λ8 (Ross 101–2).

259ᵃ13–20

Aristotle's argument seems to go as follows:

(1) Motion in the cosmos is everlasting (as 'has been demonstrated' (in Ch. 1)).
(2) Motion that is everlasting is continuous.
(3) Thus, cosmic motion is continuous.
(4) Continuous motion has a single mover and a single moved.
(5) Thus, cosmic motion has a single mover.

Point (3) is trivially true by virtue of the fact that every motion is continuous (*Phys.* V. 4, 228ᵃ20): if it were not continuous, it would not be the same motion, but two or more motions. Thus there is no particular need to invoke premiss (2), as if continuity were a special feature of everlasting motions. Aristotle has partly established (4) in his discussion of continuous motion at *Phys.* V. 4, 227ᵇ20–32 (cf. also 228ᵃ22: continuous motion is one). The factors of motion discussed in *Phys.* V. 4 do not, however, include the mover, but only the moved, the time, and the place or respect in which it is moved. But in *Phys.* V. 1 Aristotle does include the original mover as a defining factor (224ᵃ24). Hence (4) seems to be a reasonable extrapolation from Aristotle's account of the identity conditions of motions. In the present passage Aristotle does not explicitly draw conclusion (5), to which he is entitled and which would agree with his previous claim (ᵃ12) that a single mover suffices to explain cosmic motion.

Thus the present argument has a good systematic justification. But it seems possible to object to (4) that one might conceivably have a continuous motion with more than one mover. We might consider the case of a cannon-ball shot from a cannon, using contemporary physical concepts: the explosion in the cannon moves the ball upward and forward, but the force of gravity pulls it downward, while friction slows the forward motion; the result is a roughly parabolic trajectory that ends when the cannon-ball strikes the Earth. In modern physics the motion of the ball is analysed into vectors representing the different forces acting on the ball (we would need at least four). Of course, Aristotle would give a different analysis of the event, but it is not clear why an Aristotelian could not describe the flight of the cannon-ball as the product of two movers. The explosion in the cannon provides a forced motion upward (continued by air pressure from behind the ball), but the downward motion of the ball also has a mover. In accordance with VIII. 4, Aristotle seems to want to say that the ball is moved down by something; although we have noted difficulties with his argument there, if we grant him his claims, there will be some cause of the downward motion which will not, of course, be identical with the cannon. Hence it would appear that a continuous motion might have at least two movers even on an Aristotelian analysis. And this seems to be true whether the motions in question are forced (upward thrust of explosion) or natural (downward motion of heavy ball).

<h3 align="center">259ᵃ21–2</h3>

The MSS have *tas archas tōn kinountōn* ('the principles of the movers'). Ross notes no similar occurrence of the phrase. At ᵃ33 we meet *archē kinoumenōn* ('principle of things in motion'), but that phrase does not really pick up the initial expression. Ross suggests that *tōn kinountōn* is a gloss, and that *tas archas* refers to the beginning of the argument. Aristotle in fact recapitulates his whole argument from Ch. 3 up to the present here. I have followed Ross's reading in my translation.

<h3 align="center">259ᵃ22–ᵇ1</h3>

Aristotle recapitulates his argument (see previous note) as follows: ᵃ22–7, Ch. 3; ᵃ27–31, Ch. 4; ᵃ31–ᵇ1, Ch. 5.

<h3 align="center">259ᵇ1–20</h3>

This paragraph also looks back to an earlier discussion, but it is not strictly a recapitulation. At 2, 253ᵃ7–20 Aristotle broaches the subject of animal motion, but he defers the main discussion until later (253ᵃ20–1), i.e. until the present passage.

259ᵇ2–3:   The class of animate objects, i.e. objects with souls, includes plants. Cf. *An* II. 2, 413ᵃ20–ᵇ1; 3, 414ᵃ29–33.

259ᵇ6–7:   The one kind of motion which allegedly constitutes self-motion is motion in place: 2, 253ᵃ14–15. The power of locomotion is not found in all animals (sensation, not locomotion, is the defining quality of animals: *An* II. 2, 413^b2–4); it is found in some animals, *An* II. 3, 414^b16–17, but not in plants.

259^b11: 'environment' (*to periechon*): especially the surrounding air, which accounts for respiration. The term is already ancient in Aristotle's time, being attested to in Anaxagoras B2 and B14, but the verb designating the *apeiron* as a cosmic container may possibly go back to Anaximander (see A11, A14, A15), and at least is common in doxographic reports (DK iii. 348).

259^b12–13: Aristotle explains his theory of sleep and waking in *On Sleep* 3: nutrition, converted into blood, rises with the natural heat of the exhalation to the head, where one of several possible mechanisms (457^b10–458^a10) forces it down again, causing the head to nod and sleep to come on; hence sleep is an influx of heat leading to a natural recirculation (*antiperistasis*) of humours (457^b1–2). The animal awakens when the compressed heat of the nutrition prevails over the surrounding matter, and the purer blood (which properly is found in the head) is separated from the thicker blood (458^a10–12).

259^b15: 'the mover is distinct': this must refer to physical causes in the environment, as it is also understood by commentators (Simpl. 1258. 31 ff.; Philop. 890. 26 ff.; Alexander *ap.* Philop.), for it interacts with the self-mover as a whole. The 'first mover' of $^b17$, on the other hand, must be the unmoved part of the self-mover, the soul; for more on this part, see next section.

259^b16–20: The 'first mover' in the case of animals is the soul; the soul moves the body, and the soul, since it is integrally related to the body, moves with it. The body, then, is a tool for the agency of the soul.

ANIMAL MOTION

Aristotle's analysis of animal motion is important to his case for a first unmoved mover. For if, as he notes ($^b3$–6), animals initiate their own motions, then there are independent self-movers in the world, which means that there is a counter-example to the theory he is developing that all motions are dependent. Unfortunately, Aristotle's argument here is less than perspicuous. Several possible theses are suggested by his discussion. (i) No involuntary behaviour in animals is self-motion; (ii) no animal behaviour is self-motion; (iii) changes from rest to activity (and vice versa) are not self-motion; (iv) animal locomotion is not self-motion. Aristotle's example focuses on changes from rest to activity, but it may illustrate (i). Nevertheless, if Aristotle wants to argue for the stronger thesis (iv) (which he seems to introduce at ^b6–7 and return to at ^b15–16), he needs to establish (ii). How, then, are the several theses related? I offer the following as a possible representation of the argument:

(1) Factors in an animal's environment cause some of its motions.

(2) The (involuntary) motions which these factors cause do not originate in the animal.

(3) The motions with which an animal starts and stops are such involuntary motions.

(4) Thus, animals do not move themselves continuously.

(5) Thus, animals do not properly move themselves.

We can tighten the argument by noting an inference from (1) to (2). The logical connection between the propositions may be reconstructed by supplying two tacit premisses as follows:

(1a) If a factor in the environment causes motion to an animal, the motion is involuntary.

(1b) If an animal's action is involuntary, the animal does not originate it.

Propositions (1)–(1b) now entail (2).

Aristotle makes a distinction between voluntary (*hekousios*) and involuntary (*akousios*) motion in *MA* 11, 703^b3–11, corresponding to the better-known distinction between voluntary, involuntary, and non-voluntary action in rational agents: *EN* III. 1. Involuntary motions, such as heartbeats, are carried out 'without the mind commanding' (703^b7–8). Strictly speaking, animals other than humans do not have mind (*nous*), but presumably some lower level of intelligence will fulfil the requisite condition. (In fact, Aristotle does not discuss voluntary and involuntary motions in the present passage; I introduce them only in the interest of clarifying his claims.) Aristotle might wish to concentrate on involuntary animal motions because in these the animal exhibits no significant agency. In modern psychology Pavlov's success in controlling involuntary behaviour (through conditioned reflexes) seemed to suggest a corresponding way of controlling voluntary behaviour (through operant conditioning). Similarly, Aristotle might wish to understand animal activity on the model of involuntary behaviour. In any event, he focuses on a case in which, according to his theory (see on 259^{b}12–13), the intake and digestion (an involuntary motion) of food controls the sleeping and waking cycle of animals. The contribution, then, of the animal to sleeping and waking is minimal.

Since it is most clearly in waking and sleeping that the continuity of animal motion is manifested, one might be led to infer that we had proved (4), that the continuity of animal motion depends on the environment, not on voluntary motions from the animal. But the conclusion would be hasty. What of the ostensibly voluntary changes which may sometimes bring about a passage from inactivity to activity? For example, a reclining lioness passively observes a zebra. Suddenly, she rises and begins to stalk the potential prey. The change is not a reflex action or anything like it. It corresponds to what in a human might be a conscious decision. Arguably, the source of the motion is in the lioness herself: she originates her own motion, or has an *archē* in herself. Now Aristotle could, to be sure, produce a different analysis of the situation; e.g. he might claim that the difference between the reclining and the hunting lioness was the presence of hunger in the latter (her stomach clock has just gone off, an involuntary change). But the point is that such an analysis would require *argument*, and specifically argument independent of the sleeping–waking example. At most, the sleeping–waking example can show that some animal behaviours have their sources outside the animal.

Suppose, for the sake of argument, that Aristotle could in principle complete the case for (4) by showing that in all cases of changing from inactivity to activity the change is involuntary. How is (5) supposed to follow from (4)? The proper motion referred to in the statement is, as we have seen, locomotion. The fact that the animal becomes active through an involuntary process does not necessarily imply that the animal, now that it is in motion, is moving involuntarily. To make this clear, we might imagine a modern example. Scientist A claims to have produced a computer which thinks for itself and makes voluntary decisions; scientist B disputes the claim. When A gives evidence for calling the computer's actions voluntary, B replies that they are not, because A controls the on–off switch to the computer. A would be perfectly justified in replying that this fact was irrelevant: A is not claiming that the computer is self-starting or totally independent in its action, but rather that once it is running properly, it makes independent decisions. A's reply does not by itself prove A's claim, but it does show how B's objection misses the mark. In just the same way, Aristotle's argument seems irrelevant to establishing (5).

Thus far I have conceded (1) and (2), and hence thesis (i) that involuntary motions are not self-motions. But we should be sceptical about these points too. The fact that some factor from outside the animal plays a role in the animal's behaviour does not by itself prove that the related behaviour is not self-caused. A robin eats a worm, becomes drowsy, and sleeps. It is true (assuming Aristotle's account of sleep) that without the worm the robin would not have slept. But we must also consider the contribution that the robin's digestive system makes; we could put the same worm in the stomach of a herbivorous animal, and that animal would not go to sleep. To determine the outcome of ingesting a worm, we must take into account not only factors from the environment but also the response of an organism. The external factor was a necessary condition for the sleep; but so was the appropriate functioning of the digestive system (i.e. it must be functioning properly, and also be capable of digesting worms). The organism itself makes a vital contribution to the final (involuntary) behaviour. Can we not then say that the *archē* was in the animal itself? On these grounds it appears that (1b) is false.

Will such a criticism apply if the act is voluntary? We could provide a similar analysis of a voluntary action. For instance, if I decide to eat fish rather than fowl at a restaurant where I was able to see both dishes, we can say that something in the environment contributed to my decision—the respective dishes, whose images became present to my mind through perception. That, or something analogous (such as my imagination of the dishes), is a necessary condition of my deciding; but it does not disqualify my choice from being mine, nor does it show that the only significant *archē* of my action is outside me. In fact, if we grant that factors in my environment 'cause' my action (in some loose sense of 'cause'), we must say that (1a) turns out to be false, since my choosing between alternatives is a paradigmatic case of voluntary action, and certainly so for Aristotle. Perhaps we should modify (1a) to state that if the sole moving cause of the action is in the environment, it is involuntary. This would allow us room to say that if there were a significant input from the desires of the animal, etc., the action would remain voluntary. This provides a more plausible account of animal responsibility. But it does nothing, on the one hand, to save the animal from being treated merely as a motion machine in relation to involuntary motions of the sort that are prominent in the present discussion, or, on the other hand, to block an opponent from saying that in the case of voluntary motions the animal does indeed originate its own motion. For in the case of voluntary motion, antecedent involuntary motions are not sufficient to explain the new motion which the animal engages in of its own volition.

On the present analysis, then, Aristotle assumes (i) without justification, never really addresses (ii), does not provide evidence for (iii) in its full generality, and does not adequately link his discussion of (i) and whatever assumptions he makes about (ii) and (iii) with (iv). The potential counter-example to his theory stands untouched. Aristotle has not shown that animal behaviour is not self-motion, but only that it is not self-contained motion.

It is possible, however, that Aristotle intends a much weaker version of the argument, that suggested by ^{b}20–2: he may intend only to prove that an animal cannot be the cause of continuous motion. For a discussion of this reading, see on ^{b}20–2.

259^{b}20–2

'On the basis of these considerations one can be sure . . .': perhaps the point of the preceding paragraph is not (5) but (4) (see section on animal motion, under ^{b}1–20). Aristotle is not concerned to show that animals do not move themselves in some significant sense, but only to show that they are not *continuous* self-movers. Aristotle's remarks on the 'first mover' of the animal's motions at ^{b}16–20, which seemed like an afterthought, are needed to fill out the tacit argument.

    (6) Self-movers are composed of a moved part and a mover (from Ch. 5).

    (7) The mover is moved only in virtue of being embodied in the moved part.

(8) What is moved in virtue of being embodied in another is moved incidentally.

(9) Thus, the mover in a self-mover is moved only incidentally.

(10) What is moved incidentally cannot be the cause of continuous motion.

(11) Thus, self-movers cannot be the cause of continuous motion.

Proposition (10) gets us to the present sentence, and (11) expresses the more general point. But what does all this have to do with (4)? Minimally, we have seen that animals—i.e. perishable self-movers—do not in fact exhibit continuous motion. The argument in (6)–(11) provides an analysis of why they do not: because they are moved by a mover that is itself moved incidentally. But why could such a mover not cause continuous motion? We have, as it were, empirical evidence (supplied by the preceding paragraph) that it does not, but not theoretical evidence that it could not. The following paragraph will give an account of how an absolutely unmoved mover differs in its action from a moved mover. That such an account lies behind the present passage seems apparent from ^{b}27–8. Thus far, however, Aristotle has given us only inductive evidence that (10) is true. It is possible that lines ^{b}15–20 are anticipations of the later argument: they show that the only causes operative on animals *qua* self-movers are (a) factors in the environment, which enter into mutual interactions with self-movers, and (b) the soul, which is moved incidentally. Both kinds of causes are themselves subject to motion at least of a minimal kind, and hence are not absolutely unmoved.

There is, in any case, a more fundamental—or perhaps trivial— reason why an analysis of animal motion will not do to save the appearances of motion in the world: the *explanandum*, motion in the world, is a cosmic phenomenon; the alleged *explanans*, animal motion, is a biological phenomenon. It would take a major leap to get from the biological sphere (part of the sublunary world) to the level of a cosmic principle. Perhaps a more relevant question than whether animal motion *per se* offers the key to cosmic motion is whether animal motion has an analogue in astronomy. Although the question seems unimportant to modern thinkers, it was an important issue in the fourth century BC. Plato envisaged an analogue of animal motion in the stars (*Tim.* 40b: the fixed stars are 'animals divine and everlasting'; *Laws* X, 898d–899b), as apparently did the early Aristotle (Cicero, *Nat. D.* 2. 15. 42, 16. 44 = frr. 23, 24 Rose^3). Indeed, according to Plato the whole world is a living creature, *Tim.* 30a–31a. Aristotle seems to beg the question against his dialectical opponents at ^{b}28–31.

In the end, we need to return to the proposition that seems to be central to any understanding of the argument on animal motion: (4), that animals do not move themselves continuously. In one sense this is a curious conclusion to draw. For it is a presupposition of the view that Aristotle is criticizing precisely that animals do not move themselves continuously. That is, Aristotle's dialectical opponent begins by assuming that animals at one time are at rest, then initiate motion, then relapse into a state of rest; and this shows that motion can start and stop. Why would this opponent need to have Aristotle prove (4)? Perhaps we could say that Aristotle had achieved a clarification by progressing from the assumption that animals do not *move* continuously to the consequence that animals do not move *themselves* continuously. Still, (4) will hardly come as news to Aristotle's opponent. In the present paragraph Aristotle wants to proceed from the claim that animals do not move themselves continuously to the claim that they cannot cause continuous motion in the cosmos. Behind Aristotle's argument is his ever-present desire to find a grand cause of cosmic motion. Accordingly, he seems to assume that if the opponent's self-mover is not adequate as a grand cause, it is good for nothing. But Aristotle's aim will seem curiously irrelevant to his opponent, and so, consequently, will his argument. The point of the opponent's argument is precisely that we do not need a transcendent cause of continuous motion, but simply a large set of self-movers, some of which are in motion at a given time, some not. Indeed, the opponent will not even insist on strictly continuous motion, since for him consecutive motion will do.

From another point of view, Aristotle's own account of animal motion could give a critic grounds for undermining Aristotle's claim that animals do not move themselves continuously. For reasons given in connection with ^{b}1–20, we see that even involuntary motions can count as instances of self-motion; but for Aristotle's account of animal motion to be correct, there must be continuous—or rather, successive—motions going on in the animal. For example, during sleep the animal is digesting; when digestion is completed, the animal wakes and initiates locomotion. *Something* is always going on in the animal—i.e. some kind of biological motion—that on a more generous account than Aristotle's would count as self-motion. This result might not please Aristotle's immediate dialectical opponent. But it would give grounds to another kind of opponent, one who believes in animal self-motion as an instance of natural motion in the cosmos. To be sure, the animal is not a self-contained mover, but neither is anything else. What constitutes nature is a system of interacting movers that mutually influence each other: an ecosystem. For such an opponent Aristotle's argument shows the futility of seeking a completely self-sufficient source of motion; but it does not show that there is some transcendent source of motion.

259ᵇ24

'as we have said': in Ch. 1.

259ᵇ26

'and what-is itself is to remain in itself and in the same': this characteristic of being, or, more concretely, 'what-is' (*to on*), has Eleatic roots. Parmenides says of what-is (*to eon*), in almost identical language, 'It remains the same and in the same and lies by itself, and thus it stands fast there' (B8. 29–30); and even earlier, Xenophanes says of God, 'he remains ever in the same, moving not at all' (B26. 1). (What does the vague 'in the same' mean in all these passages? In the same place? In the same state? Perhaps both?) Note also that the characteristic 'unending' (*apaustos*, ᵇ25) is one that Parmenides ascribes to what-is (B8. 27) two lines before the quoted passage. Of course, Parmenides would be appalled by Aristotle using his ontological principle to defend everlasting motion. Aristotle has retained the general principle, but greatly expanded its application. He will soon apply these Eleatic properties to the unmoved mover, 260ᵃ17–19, which will show that the unmoved mover merits a kind of ontological priority, and is ultimately responsible for the Eleatic properties of the world.

259^{b}28–31

Von Arnim (44–5) thinks these lines were originally a marginal note that should have been inserted after ^{b}22, providing a phrase for them to attach to and a better continuity from ^{b}28 to ^{b}32. Note, however, that the change in text would make an awkward connection to the sentence at ^{b}22 ff. Still, the sentence seems to be less than apt wherever we put it, as though it were an added note.

259^{b}29–31:   According to Aristotle's theory of heavenly motions sketched in *Met.* Λ8, each of the heavenly bodies below the fixed stars has the poles of its sphere on an outer sphere and displaced from the pole of the outer sphere, thus producing an oblique motion. The 'principles' are presumably the intelligences associated with the several spheres.

260^{a}1–3

Strictly, the point does not prove that the first object moved by the unmoved mover is everlasting, but rather that there must be another cause besides the unmoved mover to account for earthly changes. The subsequent argument reiterates a point Aristotle makes in *GC* II. 10; *Met.* Λ5, 1071^{a}15–17; 6, 1072^{a}10–18. Aristotle actually derives his view from Plato's *Timaeus*, where the circles of the Same and the Different are required to explain heavenly motions (36b–d). On Aristotle's view the uniform rotation of the stars could explain uniform motion, but it could not explain the observed fact of coming to be and perishing in the earthly realm. Coming to be and perishing are opposite changes, and they tend to happen at different times (spring is the time of birth, autumn of death (*GC* II. 10, 336^{b}17–18), though each species has its own biological clock (ibid. 336^{b}10–15)). Some variable cause must account for seasonal variations and patterns of birth and death; this is the oblique path of the Sun, which causes it to approach in spring and to retreat in autumn. How, then, does the uniform motion of the stars get transformed into the variable motion of the Sun? In *GC* Aristotle does not attempt to answer this question, but only attributes coming to be and perishing to motion along an inclined circle (336^{a}31 ff.). His account of nested spheres in *Met.* Λ8 (see previous section) seems to provide an adequate explanation for complex orbits of the 'planets' including the Sun (the *planētai* ('wanderers') exhibit variable positions relative to the fixed stars). Since the Sun moves from an extreme northward position on the tropic of Cancer at the summer solstice to an extreme southward position on the tropic of Capricorn at the winter solstice, it occupies contrary positions of near and far, and can theoretically be responsible for different meteorological states in the sublunary regions, and hence different seasons, which immediately influence biological cycles. (Note, however, that Aristotle's story must be more complicated than this sketch suggests, for he does not hold that the Sun generates heat *per se*; its heat is the result of a large body in rapid motion near the upper atmosphere (*Cael.* II. 7, 289ᵃ19–33, *Meteor.* I. 3, esp. 340ᵇ10 ff., 341ᵃ17 ff.). It is not clear whether this account of heat generation is defensible in light of Aristotle's physics.) The outermost sphere of the fixed stars, with its uniform motion, is a remote cause of variability in nature, but not its immediate cause. Thus the circle of fixed stars properly accounts for uniform motion, the orbit of the Sun for variable change on Earth.

260ᵃ11–14

'It has become apparent from our discussions what is the solution to the problem we raised at the beginning...': the problem was introduced at 3, 253ᵃ22–4. Only now has the answer to the problem left hanging at the end of Ch. 3 become clear. There Aristotle recognized as the two remaining possibilities regarding cosmic motion (3b) that all things can move and be at rest indifferently, or (3c) that some things are always in motion, some always at rest, and some alternate between motion and rest. Now both the phenomenon and its cause emerge together: there is an unmoved mover (or several of them), a moved mover, and changeable things affected by the moved mover. We are reminded of Aristotle's claim that the scientific facts and their causes are in a sense verified at the same time (*An. Post.* II. 8, 93ᵃ16 ff.). Thus there is something absolutely unmoved, something always in motion, and things that vary between rest and motion. Consequently, (3b) is false, (3c) true.

260ᵃ17–19

'the unmoved mover . . . because it remains simple and self-identical and in the same': see on 259ᵇ26.

CHAPTER 7

260ᵃ26–8

'Since there are three kinds of motion . . .': Aristotle distinguishes
the three kinds of motion at *Phys.* V. 1, 225ᵃ34–ᵇ9. He counts six
kinds of motion in *Cat.* 14: coming to be, perishing, increase, decrease, alteration, and locomotion. In the present classification he
omits coming to be and perishing, which *Phys.* V. 1 identifies as
changes that are not motions, and combines increase and decrease
as different aspects of change in quantity. Earlier books of the
Physics do not observe the distinction between the genus change
and the species motion—see e.g. the definition of motion (*kinēsis*)
at *Phys.* III. 1, 201ᵃ9–15, with coming to be and perishing as species.
Book VIII generally follows the classification of Book V, but not
always: Ross 8 mentions as exceptions 261ᵃ4 and 264ᵇ29 (mis-
printed as ᵃ29).

260ᵃ30–2

'what increases in one sense increases by what is like . . .': see *An* II.
4, 416ᵃ21 ff., ᵇ6–7: in so far as food is raw, it is unlike, and in so far
as it is digested, it is like what it nourishes. That is, digestion
assimilates food to the body.

Growth presupposes qualitative change which makes assimila-
tion of food possible. Qualitative change, being a change from
contrary to contrary, must be effected by a cause which exemplifies
the contrary quality; e.g. a hot body causes alteration in a cool
body. But in order for the cause to affect the cool body, it must
have been brought into proximity with it. Hence qualitative change
presupposes change of place, and Aristotle has established a hierarchy of dependence among the three kinds of motion: growth →
alteration → locomotion.

260^{b}1–5

'But surely if there is alteration, there must be something causing the alteration ... this could not happen without locomotion': alteration presupposes locomotion as providing a causal condition for its activity. The argument is plausible enough, assuming that the subject does not alter itself by its own agency. But it might seem to contradict a point Aristotle made in *Cat.* 14: 'In the case of alteration there is a certain problem, whether what is altered is altered with any other type of motion. But this is not true: in virtually all our affections, or most of them, alteration happens to us without our partaking in any other motions. For neither do we have to increase in respect to the affection moved, nor do we have to decrease' (15^{a}18–27). There is not any conflict between the two passages, however. In the *Categories* Aristotle is arguing for the non-identity of the several changes by showing (focusing on alteration) that the change in question is not at the same time another kind of change. This does not preclude the possibility that one type of change has as a necessary condition of its occurring another type of change, perhaps in some other subject.

260^{b}7–15

This is a curious argument for Aristotle to make. Nowhere in his physical theory does Aristotle account for basic qualities in terms of condensation and rarefaction. Here Aristotle seems to reduce apparently simple qualities to condensation and rarefaction, and condensation and rarefaction to aggregation and segregation. But, to the contrary, Aristotle's account in *GC* II. 2 makes hot and cold, wet and dry, the basic powers; they have their own definitions (329^{b}24–32); other qualities are reducible to them (^{b}32 ff.); and they are not reducible to anything else (330^{a}25–9). The only reduction he considers is to other members of the set of four; but his failure even to consider other alternatives is significant. (Although his definitions of the hot and the cold contain the term *sunkrinein* ('aggregate' or 'combine'), they treat hot and cold as active powers of combining other things, not as themselves resulting from combination; hot is the power of combining similar things, and cold the power of combining similar and dissimilar things. The change he envisages does not seem to involve an increase of density even in the objects acted on, but only a change to a solid state or the like.)

In an argument against Empedocles and Democritus he asserts: 'The same quantity of matter does not seem to become heavier when it is compressed, but those who claim that water is contained in air and extracted from it have to maintain this view; for when water comes to be from air, it becomes heavier' (*Cael.* III. 7, 305^{b}6–10). Clearly he regards the condensation account as incompatible with his own theory, and with experience. In *GC* I. 2 he expressly argues against the theory that coming to be and perishing consist of aggregation and separation of particles: 'But unqualified and perfect coming to be and perishing cannot be defined in terms of aggregation and segregation, as some think . . . But this is precisely the source of their misunderstanding. For unqualified coming to be and perishing do not consist of aggregation and segregation, but of the whole changing from one thing to another' (317^{a}17–22). Later, in *GC* I. 5, he shows that growth cannot be accounted for in terms of aggregation of particles either. (His own theory of growth and chemical interaction is that of 'mixture', *mixis* or *krasis*, often translated 'combination'—which misleadingly suggests structured aggregation of some sort (I. 10).)

The notion that basic changes come about by aggregations and segregations is later attributed by Aristotle to Presocratic philosophers (9, 265^{b}17–266^{a}5; cf. I. 4, 187^{a}15–16), and with good reason. The terms *sunkrinesthai* and *diakrinesthai* appear as terms of art in the Presocratics (the former in Anaxagoras B4, the latter in Anaxagoras B12, B13, B17; *apokrinesthai* seems to have been a more common synonym for the latter in Anaxagoras (with different connotations), and is also found in Empedocles and Democritus, and possibly as early as Anaximander), and they are borrowed by Plato (*Laws* X, 893e4, 6–7). This suggests that the argument may be a dialectical one, as it is taken to be by Philoponus (but not Simplicius). On this reading, we must put a strong emphasis on the phrases 'seem to be' and 'are said'. Presumably Aristotle wishes to demonstrate that even advocates of alternative natural philosophies must grant this point. For more on aggregation and segregation, see below, section on 9, 265^{b}30–2.

260ᵇ16–17

'the primary ... is said in many ways': more comprehensive accounts of the various senses of 'prior' are given in *Cat.* 12 and *Met.* *Δ*11. Here not all possible senses are relevant. Aristotle's slide from talking of different senses of 'primary' to different senses of 'prior' is not an equivocation: the corresponding Greek terms *prōton* and *proteron* are related as superlative to comparative degree of the same adjective.

The threefold distinction of senses of 'primary' provides the framework for the succeeding argument: in the present paragraph Aristotle will apply the first sense, primary in order of dependence; at ᵇ29 he will examine the second sense, primary in time; and at 261ᵃ13 the third sense, primary in essence.

260ᵇ19–29

The first sense of 'primary' (cf. preceding section) concerns the relationship between dependent and independent entities: X is prior to Y if and only if Y could not exist without X but X could exist without Y; and X is primary in a class if and only if X could exist without all the other members, but they could not exist without it. This sense seems to be a fundamental metaphysical one on which Aristotle builds an ontology in the first place: primary substances are 'primary' precisely because members of other categories are dependent on them for their existence: *Cat.* 5, 2ᵇ4–6, with ᵇ15–17, ᵇ37–3ᵃ1; cf. *Met. Δ*11, 1019ᵃ2–4 (attributing the distinction to Plato); ibid. M2, 1077ᵇ2–11.

In fact, the opening paragraph of the chapter provided sufficient evidence to establish the primacy of locomotion among motions in this sense. There is, however, a distinction we could make between the ontological priority which substances have to members of other categories (the former provide the subject for the latter) and the priority which locomotion has to other motions (the former are causally necessary in some sense to the latter). But in either case, the latter could not exist without the former. The argument which Aristotle now gives does not seem to exploit the present sense to full advantage. What is Aristotle doing here? Having shown in the first chapter that there must always be motion in the cosmos, Aristotle uses this conclusion as a premiss, developing the argument as follows:

(1) There is motion continuously in the cosmos (Ch. 1).
(2) There can be motion continuously in the cosmos if either (a) motion in the cosmos is continuous or (b) motion in the cosmos is a succession of motions.
(3) Continuous motion is better than successive motion.
(4) The better always occurs in nature if it is possible.
(5) Continuous motion is possible.
(6) Thus, cosmic motion is continuous.
(7) Only locomotion is continuous.
(8) Thus, there is locomotion in the cosmos.
(9) Locomotion does not require any other kind of change.
(10) The other kinds of change require continuous locomotion.
(11) Thus, locomotion is primary.

Aristotle's use of 'continuously' as a modifier for cosmic motion is potentially misleading, since he does not intend this qualification by itself to entail continuous motion. He subsequently distinguishes the sense of how cosmic motion occurs 'continuously' from the sense of continuous motion, but then argues that cosmic motion is in fact continuous, using an axiom (4) which imputes value to nature (cf. section on 6, 259^{a}9-12 above). This of course is a weak link in the argument for a modern reader. Both points (5), as he expressly notes, and (7), which he does not, are theses which will be defended later. Point (5) specifically concerns cosmic motion: is it possible for motion in the cosmos to be continuous? Yes, Aristotle answers in Ch. 8. Point (7) will be demonstrated later in the present chapter, at 261^{a}31 ff.

Aristotle gets around to relations of dependency only in (9) and (10), where they seem almost to be an afterthought. At ^{b}26-9 he argues for (9) and (10), but the argument is so compressed that it consists virtually of the bare assertion that locomotion does not presuppose growth, alteration, coming to be, or perishing, whereas they presuppose locomotion—indeed, continuous locomotion. But why should this asymmetrical relationship hold? We might think that Aristotle is now invoking the argument in the opening paragraph of the chapter to justify the assertions here. But there Aristotle established only the priority of locomotion to other *motions*—i.e. increase/decrease and alteration; whereas here he argues for the priority of locomotion to other *changes*, i.e. the motions plus coming to be and perishing. The argument at 260ᵇ7 mentioned coming to be and perishing, but that argument was merely dialectical as far as we could see. We might supply the argument thus: coming to be presupposes growth, and growth, for reasons given at the beginning of the chapter, presupposes locomotion. But locomotion can take place without coming to be. For instance, the heavenly bodies move in place, but do not come to be.

Thus we can supply an adequate argument for (9). But what about (10)? Statement (10) says not only that other kinds of change require locomotion, but that they require *continuous* locomotion. Where has Aristotle established the need for continuous locomotion? The present argument has established the *fact* of continuous cosmic motion (at (6)), and the fact of cosmic locomotion (at (8)). We could also extrapolate to the fact of continuous cosmic locomotion. But nowhere has Aristotle argued that other kinds of change presuppose *continuous* cosmic locomotion, or even *cosmic* locomotion. That is, the only arguments we have seen so far for the dependence of other kinds of change on locomotion do not bring in cosmological considerations at all, but only what we might call the physics of change. Aristotle seems to be conflating an argument for the priority of locomotion to other changes with an argument for the continuity of cosmic change. The two arguments have different aims and require different premisses, and it is not clear that the former supports the latter or that the latter, as it stands, is relevant to the former. (10) remains unsubstantiated, and the part of this present argument which actually applies the criterion of ontological independence seems to be poorly developed at best.

All in all, it appears that Aristotle is getting ahead of himself in the argument. Two key premisses, (5) and (7), anticipate later arguments, while the points that the present argument allegedly illustrates, (9) and (10), get short shrift. It would have been better to present the three senses of 'primary' first, and to have subsumed the argument at 260ᵃ23 ff. under the first sense, at the same time expanding it to include a treatment of coming to be and perishing. In this way he could have established the ontological independence of locomotion without undue anticipation, and without falling under the suspicion of begging the question.

260ᵇ29–261ᵃ12

One might wonder why Aristotle has to argue that locomotion is primary in time. If he can show that it is primary in the ontological sense he has just examined, and in the general metaphysical sense he will next examine, that should be enough. If he must prove that locomotion is temporally prior in some sense, at least it should be sufficient to show that locomotion is prior only in one limited sense. But elsewhere—e.g. the argument for the priority of actuality vs. potentiality (*Met.* Θ8, 1049ᵇ17ff.)—Aristotle shows a propensity for philosophical overkill. Actuality is prior to potentiality in every important sense. Even if an individual is potentially an adult before it becomes an actual adult, it is produced by a previously existing adult: the chicken comes before the egg. Here too Aristotle argues for the priority in time of his favoured subject, in this case locomotion.

261ᵃ1–3: Aristotle replies that the act of generation itself presupposes a mover. In saying that the mover is not being generated, he presumably means to include the case of a parent which generates an offspring; the parent is itself generated, but not in the same act which generates the offspring. But in a remote sense the Sun also generates perishable things by governing the life cycles (*GC* II. 10; cf. on 6, 260ᵃ1–3).

261ᵃ3–5: 'It might indeed appear': the phrase actually begins with 'since', which makes no logical sense. Aristotle seems to be introducing a new objection here, but it is difficult to distinguish the present objection from the previous one.

261ᵃ5–7: 'Although this is the case for any individual thing ...': Aristotle's answer here seems to be a more metaphysical one, that one needs an everlasting mover, and then something to move that (the unmoved mover). Aristotle can make his point by extending the domain of discourse from the individual that is generated to the cosmos as a whole: the objector is thinking only of the specific case, but each particular event of generation requires a larger situation in which the event can take place: an environment in which motion is a fact. And that motion in turn presupposes everlasting cosmic motion, and ultimately an unmoved mover. However, Aristotle's present argument is so compressed that it seems to beg the question: we must spell out the priorities in question in terms of time parameters, in order to prove that locomotion is prior in time, not just prior in some general sense.

261^{a}9–11: 'none of the subsequent motions is prior . . .': the motions in question are subsequent in a biological context, where the organism cannot alter and move in place until it is born (the biological sense of *gignesthai*) and grows. Yet many animals are born able to move in place and to alter (e.g. babies become red when they cry, a case of alteration). And in fact growth, alteration, and locomotion take place in the womb, as Aristotle should know, and hence are simultaneous with, not subsequent to, coming to be. Aristotle's point is much better made in the context of ecology and cosmology than that of the biology of individual generation.

261^{a}13–17

Aristotle now turns the posteriority in generation of the individual thing to his advantage by invoking his developmental principle that what is posterior in order of generation is prior in nature (cf. *Met.* H8, 1050^{a}4–5). For the hierarchy of powers of soul, see *An* II. 3. The immobile animals referred to at ^{b}17 are what later Greek theorists called zoophytes—Aristotle does not have a generic name for them, but he perceptively recognizes their existence (e.g. *HA* I. 1, 487^{b}7 ff., VIII. 1, 588^{b}12 ff.)—animals such as the pinna and the sponge.

261^{a}20–1

'what is moved in locomotion loses its essence less than in any other kind of motion': a second argument for locomotion being prior in essence is that it is the minimal motion: other motions require a change of quality (alteration) or of quantity (increase or decrease). He could have made the point even stronger by comparing it with the changes of coming to be and perishing, both of which involve a change of substance. In modern terms we could say that since location is a relational property, change of place involves a change only of relations, not of non-relational properties. Here we see clearly that while Aristotle rejects Eleatic denials of change, he accepts Eleatic values: constancy is better than change, and the best change is the least significant change.

261ª23–6

'Quite clearly this . . .': Aristotle gives a third argument for the priority in essence of locomotion. The argument is:

(1) the self-mover moves itself properly when it moves in place;
(2) the self-mover is the principle of all things moved;
(3) hence motion in place is the primary kind of motion.

Aristotle has put forth (2) at 5, 257ª27–31. What are the grounds for (1)? They could be the two arguments of the present paragraph: (a) that the primary motion comes last in generation, and (b) that the primary motion involves the minimum change in the subject, both of which conditions are fulfilled by locomotion. The grounds for (1) may, however, be sought in the previous chapter. At 259ᵇ6–7 Aristotle argued that animals *qua* self-movers properly move themselves with only one kind of motion: namely, locomotion. If this is his justification, (1) will apply only to perishable self-movers (although it might provide inductive evidence for non-perishable self-movers); note, however, that (a) also applies only to generated self-movers strictly speaking. Whether there are any non-perishable self-movers remains to be seen, but the present argument is couched in perfectly general terms, as though it would apply to them if they existed.

261ª32–ᵇ26

'All motions and changes are from opposites to opposites . . .': Aristotle includes as species of opposites (*antikeimena*) contraries (*enantia*) and contradictories (*antiphaseis*) (*Cat.* 10 with *Int.* 6; *Met.* Δ10). At *Phys.* V. 1, 225ª34–ᵇ9 (cf. *Cat.* 14, 15ᵇ1–16) Aristotle defines changes in terms of their end-states: coming to be and perishing are changes between contradictories—i.e. from non-subject to subject and vice versa; all motions (a species of change) are between contraries, either contrary qualities (alteration) or contrary quantities (increase and decrease) or places (locomotion). The contraries relevant to place are up–down, left–right, back–forth, and the like (*Phys.* V. 5, 229ᵇ6–10, VIII. 8, 261ᵇ34–6 below; cf. *Cat.* 14, 15ᵇ4–6). When he starts out here, it is not clear whether Aristotle wants to subsume locomotions under the general principle or not; his initial examples do not include a case of locomotion. In fact, locomotion seems the most problematic case because, while there is some sense in which the contrary or contradictory end-states of all the other changes are natural, the allegedly contrary end-states of locomotion seem conventional or merely relative to the observer: e.g. left and right depend on where the observer is stationed. In any case, Aristotle quickly moves on to cases of locomotion, almost as though the initial sentence provided inductive evidence which, when generalized, could be applied to locomotion.

To be fair to Aristotle, we must note that with his doctrine of natural place, up–down, left–right, back–forth, will not be relative or arbitrary distinctions. Indeed, Aristotle inherits this sixfold scheme from Plato (*Tim.* 43b; cf. 34a, where he includes circular motion as a seventh kind). The up–down distinction is firmly grounded in Aristotle's theory of place, with down referring to the centre of the universe, up to the circumference, and heavy and light bodies moving toward the centre and the circumference of the universe, repectively (*Cael.* I. 3, I. 8). Aristotle argues further that there is a right and a left to the cosmos, the right being the east, where the stars rise (*Cael.* II. 2). He associates his three directions with the three dimensions (ibid. 284ᵇ24–5) and with the origin of motion by animate beings (ᵇ25–33), whereas we judge the motion of inanimate things relative to our own position (285ᵃ1 ff.). Thus Aristotle sees the contraries of place as having absolute reference points as well as being capable of being interpreted relative to our own position.

To anticipate Aristotle's later moves, we could query the universality of the contraries, given that circular motion will turn out to be the primary kind of motion. How is circular motion a change between contraries? Even circular motion, to be sure, admits of a distinction between clockwise and counter-clockwise. Nevertheless, motion along a circle is not defined by that distinction alone, and ultimately contraries cannot provide the defining end-points of continuous circular motion, on pain of making it non-continuous, as Aristotle himself will argue.

What, then, is Aristotle getting at here? The sentence before the
one quoted says, 'That none of the *other* kinds of motion can be
continuous is apparent from what follows' (ᵃ31). Does the following
sentence apply only to motions (strictly, changes, since coming to
be and perishing are included, as Aristotle will explicitly note at ᵇ3–
4) other than locomotion? Or perhaps only to motions other
than motions towards opposites? (That is, are locomotions along a
straight line included?) The argument itself seems to apply to all
changes between opposites, including locomotions. Thus it is probably best to understand Aristotle as intending it so to apply. The
emphasis seems to fall on the fact that motions or changes to
opposites must be finite and cannot proceed at the same time: some
period of rest must intervene. The argument goes roughly as
follows:

(1) Motions (strictly: changes, with the possible exception of
    some locomotions) are between opposites.
(2) It is impossible to move in opposite directions at the same
    time.
(3) Motion from opposite to opposite takes a finite time.
(4) Thus, a body cannot have been moving for an infinite time
    towards a given opposite.
(5) What is not always moved with the same motion must have
    been at rest previous to its present motion.
(6) When it is changing, it is not at the opposite which is its goal.
(7) Thus, it was at rest in the opposite state.
(8) Thus, its motion is not continuous.

On this analysis we can see what Aristotle is getting at. For
the argument to work, we must suppose that there is a class of
motions to which (1) does not apply. Not all changes, then, are
between opposites; nor, specifically, are all motions between contraries. This conclusion constitutes an exception to Aristotle's
stated principle that motions are between contraries (and changes
are between opposites), and marks at least a minor development in
his thought. Aristotle notes the exception at *Phys.* VI. 10, 241ᵇ2–3,
ᵇ18–20.

261ᵇ1–2: 'But what is not always moved with a certain motion, if
it existed previous to undergoing the motion, must previously have
been at rest': the rationale for this statement (statement (5) in the

foregoing analysis)—indeed, its meaning—is not at all clear. If a moving body M proceeding from contrary A to contrary B was not describing that motion at some previous time, it could have been resting; but it could equally well have been moving from B to A. Aristotle could go on to argue that before reversing direction to move toward B, M must have been at rest, as he will at the beginning of Ch. 8. But that conclusion is not obvious at this point. Ross's comment (following Simplicius) that if M had not been at rest at A, 'it must have been simultaneously moving to and from state A' seems false. In Hardie and Gaye's translation 'must previously have been at rest *so far as that motion is concerned*' the phrase italicized (by me) is pure invention. Cf. Wicksteed and Cornford's 'must then have been exempt from that movement'. Also problematic is the claim that M will have been at rest in the contrary state; it may have been at rest in an intermediate state.

There seem to be two possible strategies for establishing the claim that contrary motions are not continuous. One would be to point out that contraries are logically incompatible, so that a motion toward A cannot be identical to a motion toward B, and hence at least two motions are involved in a cycle from A to B back to A (cf. ᵇ9). At the end of the paragraph Aristotle acknowledges that the incompatibility of predicates lies at the basis of the present analysis. But this strategy would not have anything to say about a state of rest at B. The other strategy would be to analyse what happens when the subject reaches and departs from B. This strategy might yield an account according to which M rests at B. But some argument would be necessary to establish the point that M must rest at B. To the contrary, however, it can be shown that motion to contrary extremes does not by itself entail rest. Take the example of a sine wave, which fluctuates between minimum and maximum values while describing a continuous curve. A moving point on the curve need never pause in its journey between extremes.

261ᵇ3: 'And the case will be similar with changes which are not motions': the potential problem with changes which are not motions (i.e. coming to be and perishing) is that (a) they are between contradictories (being and not being) rather than between contraries, and (b) one of the goals of change, not being, is undefined. In reference to (a) Aristotle shows that it is sufficient that coming to be and perishing are opposites which cannot coexist in the same subject at the same time. On this point cf. *Met.* Δ10, 1018ᵃ22–5, where Aristotle defines opposites in terms of non-coexistence— 'things which cannot coexist in a subject receptive of both are said to be opposites'—and applies the definition to grey and white, thus showing that it applies to different determinables on a scale as well as to the contrary extremes of the scale. Indeed, it is the fact that contraries are incompatible which renders them unable to be exemplified at the same time, and hence which makes contrary motions non-continuous. And contradictories show their incompatibility more transparently than do contraries. As to (b), it creates two related sub-problems (ᵇ10–12): (i) it makes no sense to say that M rests at B if M ceases to exist at B (as happens in the case of perishing); nor (ii) can there be a state of rest contrary to the state of change, precisely because (i) is true. Aristotle circumvents the problems by finding a deeper sense in which the conditions found in contrary motions apply also to opposite changes: it does not matter whether we can use the term 'rest' or not, so long as there is a time interval between the motion to B and the return motion to A, which he assures us there is. (Since he gives no new arguments for the interval, we must draw on those already adduced for contrary motions.) Aristotle's argument here is unimpeachable: if there is an interval between contrary motions, there will be an interval between opposite changes. But alas, we have reason to doubt that the condition is satisfied for contrary motions.

At *Phys.* V. 6, 230ᵃ7–18 Aristotle discusses the problem of rest for a state of not being. He identifies a state of absence of change (*ametablēsia*) which can be present in the existent subject and asks: (i) is there a corresponding absence of change when the subject ceases to exist, and (ii) is it rest? Aristotle does not answer the first question, and poses a dilemma to the second: if it is, either (a) not every state of rest will be contrary to a motion, or (b) coming to be and perishing will be motions. Alternative (b) is of course ruled out by Aristotle's scheme of classifications in *Phys.* V. Why (a) should be unacceptable he does not say; but perhaps it is because, as in ordinary English 'rest' seems opposed to motion proper, so in Greek *ēremia* seems opposed to *kinēsis*. In any case, he rejects both alternatives. He concludes that absence of change is analogous to rest, but he remains non-committal about whether there is a state of absence of change when the subject does not exist.

261ᵇ15–22:   The thesis that 'to each of the contraries there is only one contrary and not many' (Plato, *Prt.* 332c8–9) is arrived at by Socrates through an inductive argument. He invokes it against Protagoras in the argument at *Prt.* 332a–333b to show that since temperance and wisdom have the same contrary (folly), they are identical. It could be used conversely to show that if one subject has two contraries, it is not really one but two subjects, as Plato uses it in his individuation of the parts of the soul in *Rep.* IV (436b–c and ff.). At *Met. I4*, 1055ª19–20, Aristotle seems to accept the principle: 'clearly it is not possible for there to be a plurality of contraries to one thing'. Here he implies that there are different senses of contrariety, senses which he illustrates but does not expound. Elsewhere in the *Physics* (V. 1, 224ᵇ30–5) he asserts that the intermediate can serve as an extreme of change relative to a contrary, which could apply to the example of the equal and the mean being contrary both to the excess and the deficiency (ᵇ19–20). The real inspiration for the present examples, however, seems to be the discussion of moral virtue in *EN* II. 8. There Aristotle maintains that extremes of behaviour are contrary both to each other and to the mean, using as his prime example the mathematical relation of equality to excess and deficiency (1108ᵇ15–19).

Aristotle makes an extended study of senses in which motions are contrary to each other and to states of rest at *Phys.* V. 5–6. There he recognizes motions as contraries if they proceed in opposite directions. The state of rest that is most properly contrary to a motion is rest in the state contrary to which the given motion is directed. (See previous section.)

261ᵇ22–4:   As Philoponus points out, it would be most strange (on Aristotle's physical theory) for creatures to come to be only to perish immediately. In that case nature would be aiming at non-existence rather than existence. Since nature does nothing in vain, even literal 'creatures of a day' must have some function to fulfil in the economy of nature—e.g. some insects survive as adults only for a day, to mate and lay eggs.

261ᵇ24–6:   Aristotle here seems to recognize the principle of the uniformity of nature, a principle which may go back to Democritus. See above, section on 1, 252ª32–ᵇ5.

CHAPTER 8

261^{b}28

'this is motion in a circle': the circle retains a special place in Greek geometry and cosmology because of its definitional simplicity and aesthetic elegance. (See Ballew 1979 for a study of circularity in Greek thought.) The rotational symmetry of the sphere appealed to Parmenides (B8. 42–4) and to Plato (*Tim.* 33b, 34b). Plato also associates circular motion with rational thought (*Tim.* 37a–c). But to say that all curves can somehow be derived from the straight line and the circle is to make a bold claim which needs argument. There are of course many complex curves which cannot in any rigorous sense be reduced to a combination of the straight line and the circle. Logical simplicity and aesthetic priority do not suffice to make the circle a principle, cause, or element of other types of curves. In cosmology the primacy of the circle would eventually be replaced by the primacy of the ellipse, with the circle as a limiting case under Kepler's first two laws; later, Newton's law of gravity would entail that bodies in gravitational fields could move also in parabolic paths, as projectiles on Earth do when they do not attain escape velocity.

261^{b}31–2

'straight and bounded motion': since there is no infinite magnitude (*Phys.* III. 5), there cannot be an infinitely extended straight line.

261^{b}34–6

'in respect of place, up is contrary to down . . .': see section on 7, 261^{a}32–^{b}26. above.

261^{b}36–262^{a}1

Phys. V. 4 gives the identity conditions of motion. See reconstruction in Penner (1970). In order for two motions to be identical, they would have to be motions of the same subject from the same starting-point to the same end-point at the same time. For instance, the movement of runner X from point A to point B between 12.20 and 12.24 p.m. on 1 July constitutes a single motion. That motion will not be identical to a movement by runner Y from A to B between 12.20 and 12.24 p.m. on 1 July, nor to a movement by X from point A to point C at any time, nor to a movement by X from point A to point B between 12.20 and 12.24 p.m. on 2 July. In short, each run constitutes a particular event uniquely defined by the subject, the time, and the 'respect' of the motion. Note that the 'respect' does not have to consist of places as in the example I have given: X's turning red between 12.20 and 12.24 p.m. on 1 July constitutes another Aristotelian motion, one that happens to be causally related to X's run, but not identical to it because the respect of the motion is different (and perhaps, in this case, the subject is different as well: X's skin changes colour, but X's body runs the race).

262ᵃ5–12

In order to be continuous, two motions would have to be consecutive, with the latter picking up where the former left off, with no intervening period. Clearly a motion of a subject along a straight line from A to B will not be continuous with a consecutive motion from B to A, because the termini which define the motion are in reverse order.

Contrary motion in a circle can be illustrated by a diagram. Properly the motions do not stop each other by being motions to B and C, respectively, but by continuing through those points and meeting, say, at D. According to Aristotle, the motions would destroy each other if they had equal but opposite velocities (measured at the tangent to D, where they meet); one could destroy the other if it were larger and able to

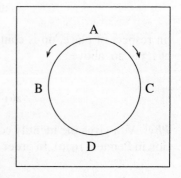

continue after displacing the slower body. In fact, according to modern physics, two elastic bodies will rebound off each other, while two inelastic bodies could cancel each other's motion, with their previous kinetic energy being converted to heat. A modern parallel to Aristotle's insight is the notion of vector addition producing a zero sum. It is exemplified in equal but opposite forces acting on a point—but not in colliding bodies.

262ª12–13

The argument from this point until 264ª7 becomes somewhat unclear. Aristotle announces that he will prove that 'motion along a straight line cannot be continuous'. He has already made that point, at 261ᵇ31–4, and indeed given the essence of his argument, which he claimed was 'clear'. But he has not really argued the point in detail, so we may perhaps indulge him if he anticipates. The real problem is that the argument following 262ª12 does not seem to follow a clear track to his intended conclusion. Along the way he introduces an extended discussion of mid-points on a line, showing that they must function only as potential points on a line crossed by a continuously moving body (ª19ff.); he shows how this analysis solves an aporia (ᵇ8ff.); and only at ᵇ21ff. does he get around to showing how the analysis does not save motion along a finite (necessarily finite, by the laws of Aristotelian physics) straight line from being non-continuous. As part of the same extended discussion he then shows how the analysis of mid-points provides a solution to one of Zeno's arguments (263ª4ff.)—indeed, a better solution than he had provided earlier in Book VI. And as a kind of appendix to the analysis of mid-points he finally shows that the dividing point of a change must belong to the end-state achieved by the process rather than to the beginning state (263ᵇ9ff.).

In this discussion it is easy to lose sight of the fact that Aristotle is ostensibly arguing that motion along a straight line must be discontinuous. The Greek commentators tend to see every reference as somehow tied to the fact that motion along a straight line must double back. But in reality Aristotle seems to become so interested in his analysis of mid-points on straight lines (with no immediate concern for the problem of doubling back) that he tends to bury his discussion of motion on a straight line in the middle of a constructive discussion. In deference to Aristotle's stated purpose at the beginning of this paragraph—namely, to prove that 'motion along a straight line cannot be continuous'—I put the whole argument under B in Appendix I; but in light of his attention to circular motion I am tempted to identify as main point B′ 'Interrupted motion is not continuous' and to promote B3 of the outline, 'This analysis solves theoretical problems about motion', to the status of main point C′. The argument that motion along a straight line is not continuous would then fall under the new C′. In any case, Aristotle accomplishes two tasks in the following argument: he shows that motion along a straight line is continuous in the short run (i.e. up to the end-point of the line), but not continuous in the long run. We must remember, too, that Aristotle's main concern is to show that straight motion in the cosmos cannot be everlasting. So we are not dealing here with pure kinematics, but with a point about motion in the real world.

262ª20–8

The cryptic remark 'the middle stands in the place of the opposite in relation to each extreme' is explained in the following discussion. A middle point, say B, lying between beginning point A and end-point C can serve as an end relative to A and a beginning relative to C; thus we could think of calling B *qua* end-point B_1 and B *qua* beginning point B_2. But B will only be an *actual* division of the line if there is an actual stop and start there.

Aristotle's account of linear continua is fundamentally different from the modern account. In modern theory, a linear magnitude is conceived as a set of points that is non-denumerably infinite, linearly ordered, dense (between any two points there is another point), and Dedekind-continuous (on this see section on 262ᵇ5–7 below). (See White 1992: ch. 1, esp. 32–3.) Roughly, a line is viewed as being constructed of points, which are, accordingly, ontologically prior to the line. Aristotle, by contrast, views the line as prior to the points. Points, with the exception of those which are real termini—e.g. ends of a line or beginnings and ends of motion—exist only potentially. Thus, whereas modern theory presupposes an actual infinity of points constituting a line, Aristotelian theory rules out that actual infinity, and allows only a potential infinity—i.e. allows that one can keep subdividing a line segment indefinitely.

262ª28–ᵇ4

Aristotle uses predications of the perfect 'tense' (Greek tense systems retain some of their earlier value as verb aspects: see Graham 1980) to explicate the motion. (Most commentators miss the verbal nuances here, though the Wicksteed and Cornford translation is sensitive to them.) For a case of continuous motion it will be false to say 'A has arrived at B' and 'A has departed from B'. The former perfect predication entails that (i) A arrived at B at some time in the past and (ii) A is still there—i.e. A is presently in a state it achieved in the past. To say 'A has departed from B' is, on the other hand, to say that (iii) A departed from B in the past and (iv) A is presently gone from B. Clearly (ii) is incompatible with (iii) and (iv), and the predications 'has arrived at B' and 'has departed from B' cannot be true of the same subject *at the same time* (ª32–ᵇ1). But it appears that Aristotle may want to say something even stronger than that the two perfect predications are incompatible. For he says, 'A is not able *either* to have arrived at *or* to have departed from point B' (ª28–9), rather than that A cannot have done *both*. Why are the statements individually false? In the first place, to say 'A has arrived at B' entails that A is present at B long enough for it to be in a state of remaining there; in the second place, to say 'A has departed from B' presupposes that A was at B long enough to establish its residence, as it were, at B.

262ᵇ5–7

'. . . making the point serve two functions': when A stops at B and then starts up again, B becomes part of two separate processes; it then has two functions, being an end-point of the prior motion and a beginning point for the posterior motion. So it is by definition two, and the motions cannot be continuous, because they are interrupted by a logical gap. In terms of our earlier distinction, $B_1 \neq B_2$; hence it is impossible for motion AB_1 (taking A now to be the beginning point: see following section) to be continuous with motion B_2C. (The phrase 'just as if one distinguished [the points] in thought' is obscure. It may mean that a possible conceptual distinction—described in the thought-experiment discussed in the section on 262^{a}20–8—is in this case realized in actuality.) Of course, there will be a time gap between the two motions also, as the previous discussion makes clear. Hence, from the standpoint either of the time period or of the course traversed, the two motions will not be continuous.

According to modern set theory, a linearly ordered, dense set which is Dedekind-continuous, if divided at a 'cut', will consist of two sets such that if the first has a final point, the second will have no beginning point, and if the second has a beginning point, the first will have no final point. Thus we cannot have a point which belongs to two different sets comprising the whole line.

262^{b}7–8

'But A is the point it has departed from': Aristotle now takes A to be not the body in motion but the beginning point of its motion. His aim is to show that B plays no defining role in the motion from A to C.

262^{b}10–21

The problem that Aristotle anticipates is closely related to the objections he has just discussed. He constructs two lines as shown in the diagram.

The problem arises from the fact that A is assumed to be retarded in its motion at B because it must arrive and depart from B at different times: it must make a pit stop. Thus, although A and D set out at the same time and with the same speed to cover equal distances, A will arrive at its destination later than D. Aristotle's exposition is awkward, to say the least. His statement that 'what set out and departed earlier must arrive earlier' does not really apply,

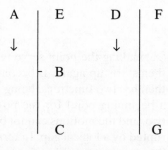

because *ex hypothesi* A and D have set out at the same time; the real difference is that A is presumed to make a stop at B, while D makes a non-stop passage. Thus A makes two departures (from E and B, respectively), while D makes only one (from F). Two sentences later Aristotle suddenly shifts from expounding the objection itself to discussing what it would take to make the objection work, given that it is incompatible with the theory he has already developed. Then he reveals what tactics we must use to block the objection. The objection can be met simply by insisting on the point which Aristotle has made already: the moving body does not occupy any point along its path for any stretch of time.

On a technical problem concerning the possibility of motions consisting of motions, see Appendix II.

262^{b}23

'if G travels up to the point where D was': Aristotle now uses 'D' for the upper extremity of line F and 'H' for a moving body starting from the lower extremity.

262^{b}23–263^{a}3

The argument seems to go like this:

(1) Any moving body M that reverses direction at a point P on a straight line must use P as an actual end and an actual beginning.

(2) Whatever uses a point as an actual end must have a time t_1 at which it has arrived at that point.

(3) Whatever uses a point as an actual beginning must have a time t_2 at which it has departed from that point.

(4) If (a) $t_1 = t_2$, (b) the moving body is at P and not at P at the same time.

(5) (4b) is impossible.

(6) Thus (4a) is false.

(7) Thus $t_1 \neq t_2$, and B is at P for a period of time.

(8) Thus M stops at P.

(9) Thus M's motion is not continuous.

Point (2) is the key premiss of the argument. Is it possible—logically possible—that M could arrive at P and not 'have arrived' at P: i.e. according to Aristotle's semantics, not be there for a time (see section on 262ᵃ28–ᵇ4)? In our experience a body reversing itself must slow down, stop, and then reverse directions, gaining speed (think e.g. of a ball thrown up in the air and falling back to Earth). Mathematically, the velocity will decrease from a positive sum to zero, then accelerate to a negative quantity. This change of velocity, however, seems to presuppose the law of inertia, which is a physical, not a logical, law. If so, the argument would need to be cast in physical terms with extra premisses. From the standpoint of logic alone, it is not clear that M needs to pause at P. To be sure, Aristotle is not claiming that the present is a purely logical argument. Still, his argument seems to based on a priori considerations of logic and perhaps metaphysics. White (1992: 60) points out that it is logically impossible for speed as well as velocity (speed and direction) to remain constant throughout a reversal, since at the extreme point speed will drop to zero. Accordingly, uniform motion does not characterize a reversal. But this point does not entail that there is a temporal break in the motion, for if the speed of the moving body is zero only at a point, then the body did not necessarily pause if it was at the extreme point of the line only for an instant of time. For a specific counter-example, see below, section on 264ᵇ1–6.

If pressed, Aristotle would probably appeal to an end-point as defining a motion (cf. *Phys.* III. 1, V. 1). The motion is completed only when M has achieved the goal of the motion and actualized its potentiality. The actualization constitutes an achieved state expressible by the perfect-tense predication of the verb of motion. M has moved, has arrived. The modern objector, however, would insist that at this point Aristotle comes close to begging the question: why should we assume that motion needs to be defined in terms of some arbitrary end-state? Why could it not continue *ad infinitum*? Something like a prejudice drawn from ordinary language analysis seems to be generating metaphysical distinctions which turn out to conflict with empirical evidence—or, more correctly, with the best principle for explaining the evidence: namely, Newton's first law of motion. (For a good discussion of the present passage, see White 1992: 54–62.)

There is one troublesome point about the argument from a systematic standpoint. According to *Phys.* VI. 5–6 there is no first moment of change for a changing body (see esp. 5, 236ᵃ14). This means, among other things, that for any given moment within the period of change, the body in motion must already have been in motion (6, 237^{a}2 ff.). This seems to suggest that we cannot find a unique time for which (3) is true. One could reply, however, that we do not need a unique t_2, merely a t_2 that is not identical to t_1. And, according to the semantics of the perfect, when it is true that M *has* arrived at P, then it is true that it has not yet departed from P; and when M has departed from P, then it is no longer true that M *has* arrived at P, but only that M *arrived* (aorist tense) at P in the past.

263^{a}4

'We must give the same kind of answer to those who raise Zeno's problem': the present analysis provides an answer to Zeno's stadium argument (cf. *Phys.* VI. 9, 239^{b}11–13).

263^{a}11

'In our earlier discussion of motion we solved the problem': Aristotle has already rebutted Zeno's argument earlier in the *Physics* (VI. 2, 233^{a}21–31). The solution is to say that time is infinite in just the same sense as the space travelled—i.e. it is infinitely divisible, but not infinitely long. Using modern tools, we could describe setting up a one-to-one correspondence between temporal moments and spatial points which would show that we will never run out of time to cross the distance travelled. Aristotle thinks in terms of stretches of time and space that can be divided on a one-to-one basis.

263^{a}15

'although this answer suffices for the question at hand': here Aristotle admits that while the earlier solution refutes the objection, it does not reveal the nature of time. Zeno's paradox depends on a correspondence between time and space continua. But we can raise the question whether time (or space) by itself is continuous. The division of a continuum into halves at a mid-point itself constitutes a threat to the integrity of the continuum. If we actually bisect line

AC at B, it becomes two lines, AB and BC, where B serves as both an end to AB and a beginning to BC. Now suppose M moves from A to C. Then M must use B as an actual end and an actual beginning. By the argument at 262ᵇ23 ff. (see above), premisses (2) and (3), M must spend some time at B. Thus M must rest at B, and consequently there must be two actual motions in the passage from A to C. Thus there will be a discontinuity in the motion just as there is a logical discontinuity between AB and BC or a logical difference between end-point B_1 and beginning point B_2 (cf. 262ᵇ5–7). To make a division actual is at the same time to destroy the continuity of a line or any other continuous quantity.

Note that in modern set theory, specifically by the property of Dedekind-continuity, if we include the mid-point of a line with the first half of the line, we cannot include it in the second half, and vice versa; we are precluded from taking the point twice, so our cut does not produce any actual gap. Aristotle makes a similar move in the next paragraph: specifically, he requires that the dividing point go with the second segment of the line.

263ᵇ9–26

The transition from one state to another, here exemplified by the change from white to not white, creates a conceptual problem of its own. When precisely does the subject become not white? Dividing the time period AB at C, the point of change, we must enquire when subject D actually becomes not white. (Aristotle refers to the time AC as A and to the time CB as B.) Since C is common to both periods, it would seem that at C, D is both white and not white. To avoid the consequence, he argues that we should assign C to stretch B and exclude it from stretch A. Whereas a modern interpreter might look at this move as a mere stipulation to avoid problems, Aristotle sees it as required by the facts themselves. Taking recourse, as he has several times previously, to the semantics of perfect predications, Aristotle regards the statement 'D has become not white' (263ᵇ22) as entailing a definite time in which D first became not white. That is, 'D has become not white' implies that at some prior time 'D became not white' is true, or, perhaps more perspicuously, 'D is not white' became true (and then remained true up to the present time). In the present case the semantics of the Greek perfect and its English counterpart are virtually the same (though not all Greek perfects behave like English perfects: Graham 1980). One can say that the point P at which the goal of the action—e.g. the predicate F—becomes a predicate of the subject S belongs logically to the end-state rather than the stage of the process in which F is becoming true but is not yet true. P marks the end of not F (or not fully F) and the beginning of F obtaining for S. From the perspective of physical-metaphysical analysis, prior to P the causal process was actual, but the result was not; from P on, the process of change is over, and the actuality resides in the result (*Phys.* III. 1; *Met.* Θ8, 1050^{a}28 ff.).

263^{b}23–4: 'So it was first true to say that it was white or not white at that time . . .': this sentence is read by Hardie and Gaye and by Ross as introducing a new alternative for the sake of completeness: namely, of D changing from not white to white. Wicksteed and Cornford avoid the unexpected alternative by translating, 'So unless we admit that a white thing can be truly described as not-white at that instant for the first time, we shall either have to say that [a] a thing does not exist at the instant when it has come to be and [b] does exist at the instant when it has ceased to be . . .' This still leaves (a) unexplained, since the example anticipates only (b). Moreover, at ^{b}28–9 Aristotle introduces another example using the same letters as before, in which D becomes white rather than not white, as if he is now contemplating both options. On either version, Aristotle seems to switch from considering one change to considering two opposite changes.

263^{b}26–264^{a}1

Aristotle sees an opportunity to turn the present argument into a refutation of the atomic view of time. The argument, however, is difficult to follow. It seems to go as follows:

(1) Suppose time is composed of atomic magnitudes, and suppose D becomes white in time AB, where A and B are consecutive atomic moments.
(2) D was becoming white in A.
(3) What is becoming F is not F.
(4) Thus, D was not white in A.

(5) D is white at B.
(6) There must be a process from not white to white.
(7) A process takes time.
(8) Thus there must be an intervening moment C between A and B.
(9) But then A and B are not consecutive atomic moments, and (1) is false.

The argument is a *reductio ad absurdum* of (1). What strikes one as odd about the argument is that (6) seems to presuppose that D is absolutely not white in A, whereas (2) says that a process was occurring in A. That is, if a process is possible in A *ex hypothesi*, why should we need another process to get from not white to white? While (3) provides a correct implication, Aristotle seems to want to assume something else: namely, that what is not F at time *t* is not *becoming* F at *t*. And clearly that would undermine the assumptions of the argument. What Aristotle *could* argue is that no process is possible in a moment, and hence the analysis will have at least to multiply moments. It may appear that (7) amounts to the claim that no process is possible at a moment; but if we interpret (7) this way, it will beg the question. For, according to Aristotle's opponent, atomic moments are parts of time—indeed, the only real parts of time. Hence they have duration—minimal duration, of course, but duration none the less. Accordingly, there is nothing contradictory about a process occurring in a moment. For Aristotle, by contrast, a moment does not have duration, but rather is an extensionless cut in the time continuum, so that (2) is incompatible with (7); but he cannot assume that analysis against his opponent. Aristotle would need to show by independent argument that a process takes time in some sense which is incompatible with atomic moments. But that seems to be a different argument from the one he is making here, one directed at (2) rather than (1).

It seems possible that Aristotle's dialectical opponents could in principle account for process in a fairly robust sense, even on the assumption that a process cannot go on within a single atomic moment. Assuming that D is a complex object, e.g. is composed of ten parts, each of which is initially not white, and that the parts can change colour independently, then in each of ten consecutive atomic moments one part of D could become white, then another. Over the whole time t_1 to t_{10} it would seem to be true that D was becoming white. To be sure, the process would not be continuous; but it would appear to be so, as in the case of a film consisting of still frames projected rapidly to produce the illusion of action. On this analysis, a process would consist of stepwise discrete minimal changes rather than continuous change; but there seems to be nothing contradictory about this analysis. By assuming that the advocate of atomic moments must compress all changes into a single moment, Aristotle imposes an unfair restriction on the opponent's theory and sets up a straw man.

But who are the mysterious advocates of atomic time? Neither Aristotle nor his commentators give any hint. An influential theory which supplied an answer was developed by Paul Tannery (1887/ 1930: 257 ff.): Parmenides and Zeno were responding to a Pythagorean theory that constructed space and time out of mathematical atoms. Although the theory was widely accepted in the first half of this century (e.g. Burnet 1892/1930, Cornford 1939, Raven 1948), it has been soundly refuted (e.g. Vlastos 1967: 376–7, although it has recently been defended by Matson 1988), eliminating the Pythagoreans as possible dialectical opponents of Aristotle (as well as of the Eleatics).

Is there anyone who constructs time out of instants to whom Aristotle is responding? Unfortunately, he never tells us. One possibility is Diodorus Cronus (dates uncertain; one firm date associated with his biography, 307 BC, would allow his life to overlap with the later Aristotle; see Furley 1967: 131) or an earlier member of the Megarian school. Diodorus is credited with the view that his principles were 'partless (*amerē*), i.e. among those things in which there are no parts' (Clem. *Strom.* 8. 15), and the view that 'some things *have moved* (*kekinēsthai*) but nothing *is in motion* (*kineisthai*)' (Stob. *Ecl.* 1. 19. 1 = Aët. 1. 23. 5). Diodorus is known to have expoited Zenonian arguments; in Diodorus's version of the arrow argument, Sextus Empiricus (*Math.* 10. 86) reports the premiss that 'a partless body ought to be contained in a partless place'. Unfortunately, it is not clear whether his destructive arguments grow out of, or support, a constructive theory of his own. At least there is evidence that one Megarian adopted (however provisionally) the concept of an indivisible place; at most we can conjecture that the Megarians might have constructed time and space out of partless places and instants. (For Diodorus as having a theory of the requisite type, see White 1992: 263–73. But White (p. 68) sees

Diodorus as responding to Aristotle rather than Aristotle to some Megarian.)

It has been argued that the Epicureans developed a geometrical theory that posited minimal parts of space (Luria 1933; see Vlastos 1965*b*: 122–5 for a history of the theory). Epicurus post-dates Aristotle, but one could attempt to connect the Epicurean theory with the earlier developments, such as theories of earlier atomists or of Diodorus Cronus or his school (see Furley 1967: 131–5). Unfortunately, however, there is no real evidence that an atomistic geometry ever existed, much less that Epicurus and his school subscribed to it (Vlastos 1965*b*, 1966; Furley 1967: 155–7). Nevertheless, the Epicureans subscribed to some sort of theory of quantum motion (Simpl. 934. 23–30). As far as one can see, though, they and Diodorus were deeply indebted to Aristotle's discussion of continuous motion in *Ph.* VI (esp. chs. 1–3); whereas there is no firm evidence that Aristotle was indebted to anyone but Zeno.

It is thus possible that Aristotle is merely developing a hypothetical position which he contemplated in Book VI (1, 231^b18–20), where he pointed out (in the broad context of Zeno's paradoxes) that anyone who tries to derive a continuum such as time from indivisibles will run into a certain type of problem. Or he may think that atomists are committed to such a view, whether they realize it or not. Bostock (1991: 208) plausibly suggests that for Aristotle time atomism generates the paradox of Zeno's arrow. Accordingly, Aristotle sees it as a theory which is presupposed by Zeno's arguments, one which he must articulate and refute in order to vindicate change. In any case, as Ross notes (p. 70), 'So far as we know, [Aristotle] was the first thinker who clearly stated the infinite divisibility of all continua.' He may also have been the first thinker to consider clearly the theoretical implications of their non-infinite divisibility.

263^b30: 'contiguous': two things are contiguous if they are 'consecutive' and 'in contact': *Phys.* V. 3, 227^a6. Cf. 264^a3, below, and 10, 267^b15.

264^a1–6

Aristotle returns to a non-atomic theory of time to show that it will escape the objection. But his solution here is odd. He says that the change is achieved only in the last moment, to which there is no consecutive moment. The time of the process together with the final moment—i.e. A plus C—do not take up more time than the process—i.e. A. His point seems to be that C is not a measurable quantity. But now it seems that he has violated his own stricture at 263ᵇ10–11 that the transition point must be attached to the final state rather than the initial process. At least theoretically, AC *is* a different time period from A alone, for it is only in AC that D *has become* white, and in AC it is no longer true that D *is becoming* white. True, the difference in duration between A and AC is not measurable, but the theoretical difference is considerable. A semantic oddity also appears here: elsewhere Aristotle seems to use the perfect tense in its usual sense of an achieved state resulting from a change. But in the present context, his perfect 'it *has* come to be' denotes only the completed action without the resultant state, the process up to and including the point of completion and closure. It would be more consistent to use the aorist predication to denote the completed event minus the continuing resultant state.

264ᵃ8

Logikōs, here translated 'general', can be used of dialectical arguments: i.e. arguments with plausible but ungrounded premisses, *Top.* I. 1, 100ᵃ29–30 (which premisses may be false: *Top.* VIII. 12, 162ᵇ27). But the term can also be applied to arguments that are general enough to apply outside a specific natural domain, without prejudice to their validity (cf. the trichotomy of questions as 'ethical', 'natural', and 'logical': *Top.* I. 14, 105ᵇ20–5). Sometimes logical arguments can be precise (*Met.* M5, 1080ᵃ9–11), though they sometimes lead to mistaken interpretations (*GA* II. 8, 747ᵇ27–30, with 748ᵃ7–16). In the present case, the more general arguments apply to all kinds of change, not just to locomotion; Aristotle accepts them as valid. The *phusikōs* or specifically natural arguments that Aristotle gives for the continuity of circular motion are found at 261ᵇ27–263ᵃ3. In the four general arguments that follow, Aristotle uses examples of locomotion twice, but only as illustrations: the arguments apply to all types of motion (cf. 264ᵃ14, 21–2, with examples of alteration in the third and fourth arguments).

264ª9–11

'Everything that moves continuously . . . was previously travelling towards the destination': this point applies only to the genuine destination of a motion; if D is in continuous motion from A to B it is not true that it sets out towards any arbitrary point C between A and B. The motion is in fact defined by its end-point (cf. *Phys.* V. 1 and V. 4), so, according to Aristotle, there is an intentionality in motion.

264ª13–14

'For why should it be moving toward its goal now rather than previously?': This argument is a descendant of Parmenides' B8. 9–10: why should something come into being at one time rather than at another? See on 1, 252ª14–16.

264ª14–20

Aristotle argues as follows:

(1) Every moving body M begins moving to its destination as soon as it sets out.
(2) Suppose that M sets out in a straight line from A to C and returns to A in a continuous motion.
(3) Then M is moving towards A as soon as it sets out from A (1, 2).
(4) Motion from C to A is contrary to motion from A to C.
(5) Thus, M is moving in contrary directions (3, 4).
(6) If M is moving towards A, it must be moving from C (2).
(7) M has not yet arrived at C.
(8) Thus M is moving from a point where it has not been (6, 7).
(9) (5) and (8) are impossible.
(10) Thus (2) is false.

In one sense this argument works quite well as an articulation of the assumptions built into Aristotle's concept of motion along a straight line. One wonders, however, if the argument might be susceptible to an Aristotelian-type distinction that could show back-and-forth motion as continuous. Applying the potentiality–

actuality distinction to the situation at hand, we might say that M is
actually travelling towards C, but potentially travelling towards A;
it is actually moving from A, but potentially moving from C. This
move would block (5) and (8), and it is not an obviously fallacious
move. Elsewhere Aristotle is more than ready to apply his
potentiality–actuality distinction to defend problematic situations
he wishes to accept.

We might compare the case of motion in a circular path from A
to mid-point C then back to A. Here, of course, we modify assumption (2), and we remove the grounds for asserting (4): motion from
C to A, as long as it continues in the same direction—e.g. clockwise—is not contrary to motion from A to C. But since it becomes
problematic just what the destination of M is on a circular path, we
may have to bring in some notion of potentiality: M is actually
moving towards C, but is potentially moving towards A. Why, then,
is the concept of potentiality–actuality not appropriate to motion in
a straight line?

Of course, the argument also depends upon (1), which is rejected
by modern physics. A body in motion has no natural destination,
but tends to stay in motion without reference to any particular
point along its path.

<h2 style="text-align:center">264^a22–4</h2>

'each kind of motion': the three sorts of motion are locomotion,
alteration, and increase and decrease (*Phys.* V. 1–2); on the notion
of rest as opposed to motion, see *Phys.* V. 6 and section on 7,
261^b15–22 above.

<h2 style="text-align:center">264^a24–7</h2>

If a body that undergoes intermittent motion starts to move, it must
previously have been at rest. This point is equivalent to the claim
that if a body departs from the starting-point of its motion, it
must have been at that point for a period of time (cf. on 262^a28–^b4).
As long as we think of motion as defined by starting and ending
points, this seems plausible. But modern physics makes no such
assumptions.

264^{b}1–6

The change from not white and the change to white constitute, to use modern terminology, not two different changes but one change under two descriptions. Hence they pose no problems, and it is a bit confusing for Aristotle to bring in the two descriptions. Presumably he does so to emphasize the incompatibility of becoming not white with what has happened: the new process contrasts with both the state of whiteness and the process from not white. But the argument does not work. Consider an example used above (section on 7, 261^{b}1–2), of contrary locomotion along a continuous sine wave: a body M moving along the wave will move alternately up and down on a continuous track. At the extreme bottom of the arc, call it P, where $y = -1$, M will be down as far as it can go. Immediately it will begin to move upward. But it does not follow that M was at rest at P: P consists of a single point which the body occupies for only an instant of time. Since rest requires extended duration, M was not at rest at P. By analogy, we could at least imagine a continuous change of colour in which the surface S achieved colour C only at an instant and then began its cycle toward the opposite colour. Continuity of change does not require that the contrary colour change begin at the exact moment of reaching the extreme colour, but only in the period following that moment.

264^{b}6–8

Aristotelian motions are defined as being from a beginning point to an end-point (*Phys.* V. 1, 225^{a}34–^{b}9, with 224^{a}35–^{b}1; cf. section on 7, 261^{a}32 above), and it is crucial that there is an asymmetry between the two extremes: we cannot interchange beginning and end-points because of the asymmetry of time. Hence, by definition, the motion from A to C is different from the motion from C to A. Since contraries are incompatible, they cannot both be e.g. C; if one is C, the other must be A.

264^{b}11

Prothesis, here translated 'act', conforms to none of the definitions in LSJ ('placing in public', 'statement', 'theme, thesis', 'purpose', 'supposition'); Bonitz 638^{a}44 records it with a question mark. I assume that a generic sense of the verb *protithēmi* lies behind the use: 'with the same act of being put forward' (cf. LSJ III). But even on this interpretation the term hardly seems apt; it is more commonly used of proposals and purposes. Themistius glosses the term with *hormē* ('impetus'), Simplicius with *hormē* and *rhopē* ('inclination'), Philoponus unhelpfully with 'circular motion'.

264^{b}13–17

Aristotle seems to have in mind a circle with a diameter drawn through it (cf. Cornford's note *ad loc.*). Contrary motions would be in opposite directions towards the extremes of the diameter, while motions in different directions along the circumference would qualify as opposite motions, but not as contrary motions. For at some points along the circumference the opposite motions would not be maximally distant from each other. Aristotle defines the contrary in place as the points that are maximally distant from each other on a straight line, i.e. the extremes (*Phys.* V. 3, 226^{a}32–3 = *Met.* K12, 1068^{b}30) in accordance with his general definition of contraries as the maximally different (*Met.* Δ10, 1018^{a}25–31).

264^{b}19–24

'Covers the same stretches . . .': Aristotle contrasts motion which is never with motion which is repeatedly *en tois autois* (literally, 'in the same things'). The meaning of the expression is obscure. Hardie and Gaye and Ross, apparently following Themistius (321. 17–19, not line 25 as Cornford says), take it to mean 'in the same limits'; Cornford, citing *Phys.* VI. 9, 240^{a}33–^{b}1, takes it to mean 'at the same points'. The former translation is difficult to reconcile with the parallel phrase *en allōi kai allōi gignomenēn* (literally, 'comes to be in ever different'), ^{b}20–1, where we would expect Aristotle to say that circular motion does *not* have limits rather than that it has ever different limits. The second translation handles ^{b}20–1 nicely, but makes it unclear how straight motion could be 'at the same points' in some sense in which circular motion is *not* 'at the same points'; for surely the moving body crosses the same points on each revolution of a circle. I believe that we must take it to mean something like 'over the same stretches', where those stretches are defined, in Aristotelian theory, by their end-points, and where the adverb 'repeatedly' carries the connotation of retracing them. If we do this, the presupposition of ^{b}23–4 that straight motion will take place in contrary directions makes sense; otherwise it does not. Cf. Wagner's translation, *ein und dasselbe Stück* ('one and the same stretch'). This interpretation seems to agree with Aristotle's treatment of traversing the middle part of a change, in the next paragraph (264^{b}30–265^{a}2).

Aquinas provides another solution which is also attractive. According to his reading, motion on the diameter of a circle from A to B must retrace its steps from B to A through the same points; but motion on the circumference along arc AB in its return to A covers new territory along the arc BA (see diagram). This solution can be combined with the one I have recommended above if we recognize that a stretch of a circle

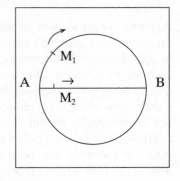

is defined not only by the end-points (A and B) but by the direction of motion (clockwise or counter-clockwise), and motion in the same direction (e.g. clockwise) from B to A does not retrace motion from A to B.

One may object that the iterated actions in the phrase 'circular motion *never* covers the same stretches, but straight motion repeatedly does' (^{b}19–21) are incompatible with this account: Aristotle seems to have in mind everlasting motion. But we can make sense of the phrase. In the context (^{b}18–19) the course is from one point to the same point, i.e. from A to A. Within that course, body M_1 moving along the circumference of the circle never retraces the same stretch, while body M_2 moving along the diameter AB will in fact retrace AB as a whole and every sub-stretch of it on returning from B to A. Thus M_2 'would have to pursue opposite motions at the same time' (^{b}23–4), i.e. in completing its journey from A back to A. Nothing requires us to presume that Aristotle is discussing everlasting motion here, and the context counts against such a reading.

COMMENTARY 264^{b}28–33

264^{b}28–33

All other changes are like locomotion along a (finite) straight line: they are between extremes on a linear scale. Thus alteration is between extremes such as black and white, increase and decrease between some maximally large size and minimally small size for the subject, and coming to be and perishing between not being and being. Accordingly, the types of arguments applied to locomotion on a straight line apply to them, and show that they fail to qualify as candidates for continuous change. Aristotle's concern for the middle section of changes seems to support the interpretation of non-continuous changes as covering the same 'stretches' (see above on 264^{b}19–24). As he points out, it does not matter into what stages we analyse the middle stretches.

265^{a}2–7

Aristotle has Heraclitus in mind as the paradigmatic philosopher of flux. Compare *Met.* A6, 987^{a}32–4, following the interpretation of Plato, *Tht.* 152d–e, 160d *et passim*; *Crat.* 439c–d, 440b–c; *Phil.* 43a, with Heraclitus B60. Ross thinks Aristotle is alluding also to Anaxagoras, citing *Phys.* I. 4, 187^{a}30, and *GC* I. 1, 314^{a}13. But this is unnecessary: when Aristotle mentions Anaxagoras, it is only to point out that he misspeaks himself ('Anaxagoras, however, mistakes his own position': *GC* I. 1, 314^{a}13), since as a pluralist he should distinguish between alteration and coming to be (ibid. ^{a}8–15).

On the question of whether Heraclitus actually held the doctrine of universal flux, see above on 3, 253^{b}9–11. However we take Heraclitus's position, Aristotle clearly takes him both as a material monist and as a philosopher of flux. He has already offered a preliminary rejection of Heraclitus's view of flux at 3, 253^{b}6 ff., where he suggests that his position is vague ('Although they do not specify what sort of motion they mean . . .': 253^{b}11–12) but none the less vulnerable to criticism. If Heraclitus were a material monist, however, he would be committed to at least one thing not changing in its nature: namely, fire. Thus he could not hold that there is universal flux, for at least at the level of basic ontology, there would be constancy. In fact, Heraclitus was not a material

monist (nor were any other early Ionians), despite an attempt by
Barnes (1979*b*: i, chs. 3–4) to resurrect the interpretation. For
material monism requires that the original stuff continue as a per-
manent substratum for change; but Heraclitus and the early
Ionians have a view according to which the original stuff does not
remain through its transformations (see Heidel 1906; Cherniss
1935: 359 ff.; Stokes 1971: 30 ff. Barnes's key arguments are anticipated by Stokes and refuted (pp. 40–1); see also Graham 1997:
7 ff.). The early Ionian *archē* is an original source, but not an under-
lying principle.

Ironically, the one type of change which Aristotle does not seem
to have effectively ruled out as potentially continuous is the one he
seems to think easiest to dispose of: coming to be and perishing. By
trying to assimilate the early Ionians' theories to his own categorial
scheme, Aristotle has overlooked the fact that what they advocate
is not alteration of a single *archē* which allegedly persists as a
substratum, but rather substantial transformation of that *archē*.
Thus for Anaximenes air becomes fire by rarefaction, and wind,
cloud, water, earth, and stones by condensation (DK A5, A7).
Although the Ionians lack a vocabulary for analysing changes, we
can say that the transformations in question are more like Aristotle's coming to be and perishing than like his alterations (in
fact they are very much like Aristotle's elemental transformations
minus an account of prime matter). Already in Anaximander
the contraries pay retribution to each other by perishing into
each other (B1), and Heraclitus thematizes the point by saying
that it is death for water to become earth, but that water comes to
be (or is born) from earth (B36; cf. B76). Life and death are surely
paradigmatic instances of changes of nature, not of accidental
properties.

Aristotle has given no reason for the Presocratics not to say that
the coming to be and perishing of the natural elements are cyclical,
as indeed Aristotle himself thinks they tend to be (*GC* II. 4, 331ᵇ2–
4), and if the processes are cyclical, why should they not be continuous? Aristotle can make the conceptual point that, by definition,
coming to be and perishing are changes between being and not
being and hence have a determinate end-point. But it is not clear
why a dissenter could not say that under another description matter
proceeds from one condition to another without arriving at any
state that constitutes a genuine resting-place. The terms 'being' and

'not being' which Aristotle wishes to impose are merely relative to
a given change—e.g. when earth changes to water, water gains
being, earth gains not being; while in the reverse process, earth
gains being, water not being. Although we can identify set points in
the cycle, we cannot say that matter rests at those points as having
achieved an absolute goal of change—certainly calling some point
'being' does not confer finality upon it. Aristotle might wish to
make the more basic (for him empirical) point that there is rest at
each stage of the cycle of transformations just because a given
element persists for a time. But if he is not to beg the question by
assuming his own physics here, he should further show why e.g.
Heraclitus would be wrong to claim that at least some portion of,
say, water is not changing at any given moment, and hence there is
continuous change going on. Aristotle might reply that Heraclitus
would be justified in saying that matter was changing continuously,
but not in claiming that continuous change was going on; for each
portion of water that changes into e.g. earth rests as earth. But what
is the evidence for that? Perhaps in the end he would appeal to our
sense experience, as he has done already in a similar context at 3,
253ᵇ28 ff.

   In fact, the argument is moot, because the early Ionians do
not seem to recognize cyclical but only sequential becoming
(Anaximenes, DK A5, A7; Heraclitus B31, B30, B60: natural
change is not a cycle but a two-way street supporting 'up' and
'down' motions, with termini at either end). Nevertheless, it does
not appear that Aristotle has adequately forestalled an argument
for continuous cyclical generation of elements.

CHAPTER 9

265ª13–15

See 8, 261ᵇ28–31, with commentary. By making circular and
straight motion the elements of all other motions, Aristotle obvi-
ates the need to examine complex motions separately: only the
properties of straight and circular motions need be considered. But
as we have noted (on 261ᵇ28–9), it is as impossible to reduce all
motions to straight and circular motions as it is to reduce all figures
to straight lines and circles.

265ª16–24

265ª16–17:   'it is simple and more complete': the simplicity in question should not be confused with the simplicity of the straight and the circular *vis-à-vis* a composite motion: in that context both are equally simple. The sense of 'simple' emerges from the context. Here Aristotle sets up a complex dilemma:

  (1) What is simple and complete is prior to what is not simple and complete.
  (2) Motion along a straight line is along either (a) an infinite or (b) a finite line.
  (3) There is no infinite line.
  (4) Thus (2a) is impossible.
  (5) Motion along a finite line must either (i) double back or (ii) not.
  (6) If (i), the motion is not simple, but a complex of contrary motions.
  (7) If (ii), the motion is not complete.
  (8) Thus, motion along a straight line is not simple or complete.
  (9) Motion in a circle is simple and complete.
  (10) Thus, motion in a circle is prior to motion along a straight line.

The relevant sense of 'simple' becomes clear in (6). The argument effectively employs Aristotelian distinctions to show the superiority of circular motion.

265ª17–20:   'It is not possible to traverse an infinite straight line . . .': here Aristotle gives two distinct reasons why motion along a straight line cannot be ultimate: (a) there is no infinite straight line, and (b) even if there were, it could not be crossed. The first point is argued in *Phys.* III. 5, where the arguments are directed primarily towards the physical impossibility of an infinite body, of which an infinite quantity would be the extension. The second point is justified briefly here. Compare Aristotle's arguments in *Phys.* VI. 2, 233ª31 ff., and 7, 238ª20 ff., that a motion over an infinite distance cannot be realized in a finite time. This argument is basically an argument based on kinematics, in which time and space and the moving body are related in an abstract theoretical way. In *Cael.* I. 5–7 Aristotle also gives arguments against the possibility of an infinite body; his arguments there bring in a mixture of cosmological, physical, and kinematic considerations.

265ᵃ22–3:   'For the complete is prior to the incomplete, in nature, in definition, and in time': the distinction of the senses of priority— (a) in nature, (b) in definition, (c) in time—can be compared with the distinction at 7, 260ᵇ17–19: priority (i) in dependence relations, (ii) in time, and (iii) in essence. Despite Simplicius's attempt to reconcile the schemes (1314. 19ff.), only priority in time appears to be the same. Neither (i) nor (iii) seems to be reducible to (b) definition. Against Simplicius's identification of (a) and (i), Ross rightly notes that Aristotle uses 'in nature' in reference rather to (iii) at 261ᵃ14. Ross also notes an interesting parallel between (a), (b), and (c) and Aristotle's discussion of the priority of actuality to potentiality at *Met. Θ*8, 1049ᵇ10ff.

How shall we account for the difference? What Simplicius and Ross do not notice is that the subject of (i)–(iii) is the general— one might say the ontological—priority of kinds of motion; but the subject of (a)–(c) is the priority of the complete (*teleion*) to the incomplete (265ᵃ23). Hence, to try to correlate the two schemes is to compare apples with oranges. Ross rightly makes use of the notion of the complete in reference to (b) definition: one defines the incomplete in terms of the complete. Similarly, (a) the paradigmatic object in nature is the complete object; e.g. the mature oak-tree is prior by nature to the acorn. And (c) in time the complete is prior to the incomplete, because only a complete specimen can produce an incomplete one; e.g. only an oak can produce an acorn: the chicken always precedes the egg on Aristotelian assumptions (*Met. Θ*8, 1049ᵇ17ff.). The present argument finds a nice parallel in *Met. Θ*8 precisely because the notions of completeness (cf. *entelecheia*) and actuality (*energeia*) (related at *Met. Θ*1, 1045ᵇ33–1046ᵃ2, 3, 1047ᵃ30–2; see Graham 1989) overlap significantly.

265ᵃ24–7

See *Met. Θ*8, 1050ᵇ6ff. on the priority of the everlasting. The relation of this point to the previous argument (ᵃ16–24) is not clear. It could be meant as a corollary of the fact that circular motion is simple and complete; or it could be a totally independent point.

What seems curious is that this argument appears almost as an incidental afterthought; yet it should be a pivotal argument, for it is the capacity of circular motion to be everlasting that will allow Aristotle to apply the unity and continuity of motion, proved in Ch. 8, to explain everlasting cosmic motion, the *explanandum* introduced in Ch. 1. Note, however, that the present argument will not by itself take us all the way to an explanation of the *explanandum*: rather than saying that circular motion *is* everlasting, Aristotle says only that it *can* be everlasting. It will take an additional causal account to show that some circular motion in the cosmos in fact is everlasting. Still, the deceptively modest present argument marks an important step in the overall proof of Book VIII.

265ᵃ27–ᵇ8

Aristotle argues here for the continuity of circular motion as contrasted with motion along a straight line.

  (A) Motion in a straight line is not continuous (ᵃ29–32).
    (1) Any motion that has limits rests at those limits.
    (2) Motion in a straight line has determinate beginning, middle, and end points.
    (3) Beginning, middle, and end points are limits.
    (4) Thus, motion in a straight line must come to rest.
  (B) Circular motion is continuous (ᵃ32–ᵇ8).
    (1) On a circle any point can be viewed as beginning, middle, or end.
    (2) Thus no point on a circle is really a beginning, middle, or end.
    (3) Thus a circle is unlimited.
    (4) The centre is really the beginning, middle, and end.
    (5) The centre is not on the circumference.
    (6) Thus, the moving body does not pass through the beginning, middle, or end point.
    (7) Thus there is no place for the moving body to rest.
    (8) Thus circular motion is continuous.

Argument (A) continues the assumptions and arguments of the present book and previous books of the *Physics*—especially Book V, which provides the identity conditions of changes. The reference to a middle seems less apt than the references to a beginning and an end: the middle point (or stretch) does not really serve as a limit, but at most, as we have seen (8, 262ᵃ19ff.; see above on 262ᵃ12–28), as a potential limit. The middle does play a role in Aristotle's consideration about how motion doubles back on a straight line (8, 264ᵇ19ff.); but that particular consideration is not illuminated by the present argument. Aristotle may bring in the middle point simply in order to contrast it with the middle point of the circle.

Argument (B) stresses the differences between circular and straight motion. (B3), expressed at ᵃ32, blocks the application of (A1) to circular motion. The contrast between (B1) and (B2) can be put in terms of the actuality–potentiality distinction: every point is potentially e.g. a beginning, but precisely because of this fact, no point is actually a beginning. (B1)–(B2) can also be seen as an argument relying on the Principle of Sufficient Reason: since there is no more reason for one point to be taken to be the beginning point than another, no point is a natural or genuine beginning point. One is reminded of Heraclitus's observation that on a circle, beginning and end are common (B103). (B4) depends on a certain mathematical analysis: a circle is defined as a locus of points equidistant from the centre, which thus becomes the beginning point for the generation (Philop. 849. 8) or the definition of a circle (Simpl. 1316. 13; Them. 233. 1). And since the radii converge at the centre, the centre serves as the end to which the circle contracts (Philoponus) or at which the radii terminate (Simplicius, Themistius). Finally, the centre is spatially in the middle of the figure. We are tempted to object that these senses of beginning, middle, and end are equivocal. Aristotle would presumably reply that they are the only senses of the terms which could be meaningfully applied to a circle. Indeed, it is in accordance with modern mathematics to define curves in terms of points or lines lying outside the curve itself, e.g. the asymptote of a hyperbola or the foci of an ellipse. In just the same sense we can say that a circle is defined by its centre point and its radius; and the centre point is the beginning and end of the radius and the middle point of the curve. From (B4) and (B5) it would follow that there is no defining point on the circumference for the moving body to pass through; i.e. an assertion equivalent to (B6) could be made.

265^b8–11

From the standpoint of measures, something is primary if and only if it constitutes a standard measure. Since circular motions in the cosmos (daily motion of the heavens = daily rotation of the Earth on its axis; cycle of the Moon = revolution of the Moon about the Earth; annual course of the Sun through the Zodiac = revolution of the Earth about the Sun) provide the basic time measurements (day, month, year), they are primary. The present argument is inserted rather abruptly in the text. One is tempted to read *sumbainei* (here rendered by the noun 'connection') as marking a conclusion that 'follows convertibly' (*antistrophōs*) from the previous discussion; but the present lines do not follow from the earlier argument.

265^b11–14

As Ross notes, this point is probably not a recognition of empirical evidence that bodies accelerate from rest, but an a priori principle of attraction to their natural places. The present account adds an important dimension to our understanding of natural straight motion in Aristotle, which we find out here is non-uniform—in conformity with the phenomena. Note that Newtonian physics gives precisely the opposite interpretation of straight and circular motion: motion in a straight line is uniform and regular; motion in a circle is accelerated and presupposes the action of some external force. But in a sense, the contrast with Newtonian physics reinforces Aristotle's point: whatever motion is taken to be uniform is judged to be primary.

265^b19–22

In accordance with his best dialectical method, Aristotle brings forth the views of philosophical authorities not to prove his point, but to confirm it after he has proved it on independent philosophical grounds. Note, however, that what the authorities confirm is not the specific point that Aristotle has been working on in the last two chapters, the primacy of circular motion, but rather the weaker and more general thesis of Ch. 7, the primacy of locomotion. The authorities would not, of course, support the stronger claim: for the Presocratics, typically, circular motion is the result of a vortex, i.e. a secondary or derived motion. His first allusion is to Empedocles (cf. B8, B17. 7–8). Elsewhere Aristotle criticizes Empedocles' account of the influence of Love and Strife as inconsistent. On this, see above on 1, 252^{a}28. But in any case we can look at these two powers (somewhat anachronistically) as mere personifications of the mechanical forces of attraction and repulsion, respectively, and hence as efficient causes of locomotion resulting in 'aggregation' (*sunkrisis*) or 'segregation' (*diakrisis*) of unlike substances. On aggregation and segregation, see above on 7, 260^{b}7–15, and below on 265^{b}30–2.

265^{b}22–3

'And Anaxagoras maintains that Mind segregates things as the first cause of motion': cf. Anaxagoras B12.

265^{b}23–9

The atomists provide perhaps the most explicit example of locomotion as the principle of all cosmic interaction, since the atoms cannot do anything but move in place to arrange and rearrange themselves. Nevertheless, the motion of atoms is not strictly speaking motion in place on the Aristotelian model, for a substance can be in place only by being surrounded by matter (cf. *Phys.* IV. 4, 212^{a}20–1); hence the motion of atoms is at best analogous to a change of 'place'—which explains Aristotle's qualification, 'as it were, motion in place'.

As Aristotle notes, the atomists do not specify an efficient cause—presumably because atomic motion is underived (*Cael.* III. 2, 300^{b}8–11: they do not give an explicit account of primitive motion; see KRS 423–5 with testimonies). The closest we can come to a 'cause' is the void, which is difficult to classify anywhere in Aristotle's four-cause scheme, although it clearly is a cause by virtue of providing an explanation (on cause, see above on 1, 252^{b}4). (Although the atomists have no commitment to Aristotle's scheme of causes, Aristotle represents the typology as exhaustive; hence he must be able to accommodate everyone else's alleged causes within his scheme, even if they turn out to be erroneous attributions. The void provides the opportunity for atoms to move by being the absence of any material impediment; one might accordingly classify it as the privation of a material cause of stopping. As this negative classification indicates, however, the void is not a cause in any substantive sense, and an atomist would be well-advised to stand fast by the assumption that atomic motion (like uniform rectilinear motion in Newtonian physics) needs no explanation.

Aristotle is quite right to say that all changes derive from locomotions for the atomists; they clearly make locomotion primary.

<h2 style="text-align:center">265ᵇ30–2</h2>

'The same holds for all who derive coming to be and perishing from condensation and rarefaction . . .': the only philosopher who explicitly embraces this position is Anaximenes. Cf. Simpl. *Phys.* 24. 26–31; Ps.-Plut. *Strom.* 3; Hippol. *Haer.* 1. 7. 3. Aristotle recognizes only two alternative types of Presocratic theory concerning matter: material monism, in which one *archē* undergoes changes by condensation and rarefaction to became all other substances (*Phys.* I. 4, 187ᵃ12–16), and pluralism, in which a plurality of substances separates out of an original mixture (ibid. ᵃ20–3). Aristotle explicitly puts Anaximander in the latter class, presumably leaving the other Milesians, Thales and Anaximenes, in the former. Heraclitus is (implausibly) interpreted as accepting rarefaction and condensation by Theophrastus (*ap.* Simpl. *Phys.* 23. 33–24. 3) and the doxographical tradition (Diog. Laert. 9. 8).

Aristotle is right to recognize two different schools of thought concerning how to approach matter in the Presocratics. His specific analysis, however, is wrong: the early Ionians are not material monists as he claims, since they do not seem to recognize a *persisting substrate* for change (see above on 8, 265ᵃ2–7). Meanwhile, the alleged pluralists really are pluralists; what makes them different from the early Ionians, however, is not their being pluralists as opposed to monists, but their holding that the plural elementary bodies, whatever they are (the four roots of Empedocles, the stuffs of Anaxagoras, atoms), have an unchanging nature: the ultimate entities cannot be transformed into one another. It is only with the pluralists that the language and concepts of aggregation and segregation arise: for them, apparent coming to be and perishing just consist in aggregation and segregation of unchanging elements— i.e. of elements manifesting the properties of Eleatic Being (Anaxagoras B17, Empedocles B8–12). (For the history of aggregation–segregation terminology and also discussion of condensation–rarefaction, see above on 7, 260^{b}7–15.)

To apply the concepts of aggregation and segregation to the early Ionians is to confuse two systems of thought that Aristotle wants to keep separate. For the pluralists, as we have noted, there is no coming to be and perishing. For the early Ionians, by contrast, there *is* coming to be and perishing, which is the kind of change by which the *archē* generates and/or maintains a cosmos, and which may (e.g. in the case of Anaximenes) be driven by condensation and rarefaction—i.e. changes of concentration. The difference between the two systems, of course, derives from Parmenides' criticism of coming to be and perishing: aggregation and segregation constitute an alternative model for explaining cosmic change, one which avoids positing transformations of substance. Incidentally, Parmenides also attacks condensation and rarefaction (B8. 23–4, B8. 44–8), rendering them obsolete as explanatory concepts, and also locomotion (B8. 26–31, 41), which, as Aristotle recognizes, is a necessary presupposition of aggregation and segregation; this last challenge the pluralists do not seem to recognize or respond to.

Although variations in density do presuppose some sort of spatial redistribution of matter, this case seems the farthest removed from Aristotelian motion in place. In what sense can e.g. a quantity of air be said to have a place? Its boundaries are indefinite, and any limits identified might well be arbitrary. For problems in the theory of changing density, see *Cael.* III. 7, 305^{b}6–10.

265^{b}32–266^{a}1

'there are those who make the soul the cause of motion': this is the view of the later Plato: *Phdr.* 245c–e; *Laws* X, 895a–b.

266^{a}1–5

'And we say of what is in motion in place that it is in motion in the chief sense . . .': this assertion seems to be in part based on ordinary language analysis: we use the terms 'motion' and 'move' chiefly of cases of locomotion, and only by extension do we consider certain other kinds of changes to be motions (in Greek as in English). The present *endoxon*, or authoritative opinion, is based not on the authority of philosophers, but on the evidence of human speech embodying universal experience (see above on 3, 254ᵃ30–ᵇ1; *Top.* I. 1, 100ᵇ21–3; *Int.* 1; *An. Post.* II. 19).

266ᵃ6–9

Nearing the completion of his remarkably sustained argument, Aristotle recapitulates. 'That there always was and always will be motion': Chs. 1–2; 'what is the principle of everlasting motion': Chs. 4–5; 'what is the primary motion': Ch. 7; 'what kind of motion alone can be everlasting': Chs. 8–9; 'that the first mover is unmoved': Chs. 5–6.

Concluding Remarks

Aristotle has sought to prove that circular motion is continuous, primary in several senses, uniform, and capable of being everlasting. But there is a troublesome sense in which it seems not to be motion. If motion is by a subject for a time from one point to another, how is everlasting circular motion motion at all? For it does not proceed toward any set destination; nor does it originate from some set beginning point. One argument against the priority of straight motion is that even if there were an infinitely long straight line, it could not be traversed by any moving body. But why should we not raise an analogous objection to everlasting circular motion, that it cannot ever arrive at a destination? But if a motion is inherently unfulfillable, it is undefined, and cannot be a motion at all.

Aristotle has a deep-seated distrust of infinity. He will allow a line to be potentially divisible *ad infinitum*, but not actually to be so divided. He rejects an infinitely extended universe, an infinitely large body, and infinite space. He seems to reject infinite series of all kinds. But he allows, and indeed endorses, infinite time. With it comes an infinite series of causal events, including, presumably, an infinite chain of efficient causes. And motion turns out to be everlasting—i.e. infinite. Moreover, at least one particular motion proves to be everlasting: namely (though he does not identify it in the present book) that of the outermost sphere of the heavens. But if Aristotle is forced to admit actual infinities in some realms, why should he not admit them in others? Perhaps, we might speculate, he admits infinity only to preclude disorder. By extending time infinitely backward (and forward), Aristotle precludes the need to account for the origin (or end) of the cosmos, and with it the possibility of an infinite time period in which nothing happened, or, alternatively, in which no orderly motion was present. In other words, Aristotle extends time, which in his theory is indissolubly linked to orderly heavenly motion, back to infinity so as to assure an everlastingly ordered cosmos. His primary motion is infinite, but it is also orderly: uniform, repeatable, and confined to a closed track. Space cannot be extended infinitely, because it would then fall beyond the pale of orderly circular motion: as the radius of the universe expanded to infinity, so would the circumference, making (repeatable) circular motion impossible. Chaos would reign in the boundless, as apparently it had for many of the Presocratics. Aristotle's acceptance of infinite time and motion is not perfectly argued, but it does show a pattern of preferences in which the uniform and the regular are given pride of place against the random and the irregular. Cosmos is prior to chaos in every sense.

CHAPTER 10

266ᵃ15–23

In accordance with ᵃ12–13, we must take A to be a finite body, i.e. a body with finite size and weight. The difficult argument seems to go as follows:

(1) A moves B in an infinite time C (given).
(2) Suppose D, a part of A, moves E, a part of B, for time F.
(3) Greater motion takes more time.
(4) Motion ABC is greater than motion DEF.
(5) Thus, F is less than C (3, 4).
(6) Thus F is finite (5).
(7) As D + an increment δ and E + increment δ' approach A and B, respectively, F increases by some finite δ''.

(8) Thus F remains finite (6, 7).

(9) Thus A moves B in a finite time (7, 8).

There are a number of difficulties here. One problem emerges in (3): what precisely is a greater motion? Is it (i) motion caused *by* a greater body, (ii) motion being moved *of* a greater body, or (iii) motion over a greater distance, or all three, or something else? Concerning (i) a further question arises: what do we mean by a greater moving body? Simplicius understands the argument to be about a greater force; Ross raises problems for this interpretation, and suggests rather a body of greater size. In fact, it appears that for Aristotle the force must be proportionate to the size of the moving body if we are to make sense of his notion that we can take away a part of A and then add to it the remaining parts of A to reconstitute A. Returning, then, to (i) with this correlation in mind, it cannot be that (i) constitutes a greater motion, because the greater A is, the shorter C should be.

What of (ii)? If A remains constant and B is increased, it will take longer for A to move B over a finite distance. But it will not take an infinite time. And in any case, it would seem strange to say that A's moving B + some δ is a greater motion than A's moving B over the same distance: although it would take more effort (force) and manifest more power, the *motion* would seem to be the same.

By contrast, it does appear that (iii), for A to move B over distance S + δ does constitute a greater motion than for A to move B over S. And all things being equal, it should take A longer to move B over S + δ than it does to move it over S. But when we look at point (4), the premiss Aristotle needs to deduce (5), we find that motion ABC is compared to DEF—i.e. the motion of A moving B in time C with the motion of D moving E in time F. Nothing is said about the distance travelled by B or E, as if that were not a relevant factor. Assuming that the speeds of B and E are proportionate to the ratios of their sizes to A and D, respectively, the only way to account for the difference between C and F would be to assume that the ratio of D to E is larger than that of A to B, as Simplicius does (1322. 8-14). But all that this would accomplish would be to show that E moves over the same ground faster than B; DEF will then be a 'greater motion' than ABC in the specific sense of manifesting greater speed, in contradiction of (4). Hence, Aristotle cannot be staging a race. Perhaps he has in mind an endurance contest. Yet, even if one were to hold the speed constant, one would expect that the greater ratio of D to E would make E persist in its course longer than B does on its course, and in that sense to undergo a greater motion. In any case, there is no evidence in the text for Simplicius's interpretation.

To return to (4), why is motion ABC greater than motion DEF? We cannot say that (4) is true by virtue of C being greater than F, for that is just the point to be deduced from (4). We cannot appeal to the size of A or B, for that would be to return to (i) and (ii), which have been rejected. Is it then a combination of (i) and (ii)? It does not appear that a combination can offer any resources beyond what (i) and (ii) offer individually. As for (iii), a difference of distance can hardly come into play without some explicit reference to such a difference by Aristotle.

One conceptually different possibility, suggested by the strange phrase 'use up' (ᵃ19), is that Aristotle has in mind subtracting D from A, then subtracting, say δ_1, from A and adding it to D, then successively $\delta_2 \ldots \delta_n$, so that finally $D + \delta_1 \ldots \delta_n = A$. Similarly, we subtract E from B and then successive increments which summed with E will be equal to B. We determine then that D moves E in time t_1, that the first increment of A moves the first increment of B in time t_2, and so on. We sum all the times, which together constitute F, and compare F to C. Since F is a sum of finite numbers, F is finite, in contrast to the infinite C. This train of thought does seem to make sense of Aristotle's words, and it shows how we might be supposed to arrive at F. But of course it involves a false assumption that there must be some sort of proportionate relationship between the time it takes the whole of A to move the whole of B and the time it takes the respective parts of A to move the parts of B. Moreover, it is simply not clear that the parts of A move the parts of B at all. Aristotle has already shown that he is aware of the fallacy of division, at 3, 253ᵇ14 ff. (see comment), so it is curious that he takes no notice of the problem here—if the present interpretation is right.

There seem to remain no rational grounds for asserting (4). It appears that Aristotle has confused difference in size or power with difference in motion. (5) does not follow. The inference to (6) depends on an assumption that any number smaller than an infinite number is finite. The assumption seems plausible enough, but the modern study of transfinite numbers reveals that it is false.

Point (7) seems to embody the confusions already noted. Why should F increase as we add increments to D and E, respectively? The method reminds us of that of the infinitesimal calculus, examining what happens as we approach a limiting case incrementally. But the result seems to have no logical connection with the operation. Why should F increase if both the moving body and the moved body, or both the moving force and the resisting mass, are increasing? If motion is in some sense the result of a ratio of power between mover and moved, and that ratio remains the same, the consequent motion (whether measured by speed or distance) should remain the same. Of course, the whole thought-experiment is counterfactual for Aristotle, but the hypothesis should proceed according to some recognizable logic.

The conclusion (9) contradicts (1), and hence is meant to show that the assumption that a finite body can cause infinite motion is untenable. But the argument seems to embody serious confusions.

266ᵃ23

'that a finite magnitude cannot cause motion': strictly speaking, the subject that causes motion is not a magnitude, which is a quantity predicated of a substance—the true subject of all power and action. But throughout this chapter Aristotle seems to express himself loosely in making finite or infinite magnitude the subject of the power exhibited in the substance. (In the present sentence the word for magnitude, *megethos*, does not appear; but I read it back into the unexpressed subect from the following sentence, which clarifies Aristotle's meaning.)

266ᵃ24–6

The present argument is intended as a more general statement of the point made in the first argument: that a finite cause cannot act for an infinite time. Here Aristotle proves that a body of finite size cannot have infinite power.

266ª26–ᵇ6

Suppose a finite body with infinite power moves a given weight 10 feet in 1 second. Let some finite body with finite power be able to move the same weight the same distance in a longer period, say in 1,000 seconds. That is, time A = 1, time AB (or $A + B$) = 1,000. The power of the first body is to the power of the second as the inverse ratio of 1 to 1,000: i.e. 1,000 to 1. But if the power of the second is a finite value P, the power of the first is also a finite value: namely, 1,000P. Hence the assumption that the first body has an infinite power is contradicted. The relationship between power, distance, time, and weight is articulated at *Phys.* VII. 5, 249ᵇ27ff. The relation can be stated formally as $P = wd/t$, where 'P' represents power, 'w' weight, 'd' distance, and 't' time. This formula comes close to the modern formula for power in physics (work divided by time, where work = force × distance); for this reason (and because of its role as the property of a subject) I prefer to translate *dunamis* as 'power' rather than 'force'. Aristotle's term *ischus* in the present passage (266ª32) seems to approach the sense of 'force'. The terms, however, do not seem to function as rigid technical terms, as their apparent synonymy at *Phys.* VII. 5, 250ª4–7, indicates.

266ᵇ6–7

'Neither indeed can a finite power exist in an infinite magnitude': this whole discussion should be understood to be qualified by *per impossibile*: there can be no infinite magnitude (*Phys.* III. 5; *Cael.* I. 5–7).

266ᵇ7–8

'it is possible for a greater power to reside in a lesser magnitude': a smaller body may have more power than a larger one if the bodies are of a different type. For example, a smaller volume of earth will have more power to go down than a larger volume of water (according to Aristotle). But, Aristotle replies, a larger volume of the same substance, water in our example, will indeed have more power to go down than a smaller volume of the same substance, and that is the relevant comparison.

266ᵇ8–20

BC, a finite portion of AB, moves D in time EF. Then 2BC will move D in $\frac{1}{2}$EF, and 2^nBC will move it in $\frac{1}{2^n}$EF. As the time approaches as close to zero as we wish, we never exhaust AB, which is infinite. Hence the power, as measured by the work divided by the time (see above on 266ᵃ26–ᵇ6) is infinite.

266ᵇ15–20:   An alternative manuscript reading at ᵇ15–16—indeed, the vulgate reading—yields the following: 'Moreover, the time occupied by the action of any finite force must also be finite . . . But a force must always be infinite—just as a number or a magnitude is— if it exceeds all definite limits' (Hardie and Gaye). According to this argument, the time taken by BC and its multiples should be finite, contrary to what is the case, since there are no limits to how small the time becomes. As Ross notes, however, the transition (*de*) to the alleged new argument is too weak. Furthermore, the last sentence is a *non sequitur*: we need a link between finite time and infinite power. Wicksteed and Cornford translate the last sentence (ᵇ19–20) 'but an unlimited force (as with an unlimited number or size) must exceed any limited force', which Wicksteed ekes out with the note: 'And unless the force could increase above any assignable limit the time could not decrease below any assignable limit.' But the translation is forced, and in any case tends to make the argument (even as eked out) sound question-begging.

266ᵇ20–4

This argument is merely sketched. We may fill it out as follows: let B_1 be a body with infinite size S_1 and finite power P_1, and B_2 be a body with finite size S_2 and finite power P_2 which is a proper fraction of P_1—one that will exactly measure (*katametrēsei*) P_1— say, $\frac{1}{n}$. Then $nP_2 = P_1$, and a body of the given substance of size nS_2 will have power P_1. But since by hypothesis, S_2 is finite, nS_2 is finite, and a finite body will have the same power as an infinite body of the same substance, which is absurd.

266ᵇ28–9

'If everything in motion is moved by something, as for those that do not move themselves': that all objects in motion are moved by something is proved in Ch. 4; his making an exception of self-movers seems curious in light of the discussion at 5, 267ᵃ31 ff.

266ᵇ29–267ᵃ2

According to Aristotle's mechanics, every motion requires a cause. Forced motion requires a cause that propels the moved body, and propels it at every moment. For Aristotle does not recognize anything like Newton's first law, that a body in motion tends to stay in motion. In the case of projectile motion we may be tempted to say that the agent which throws, shoots, etc. the projectile also moves the air (or other medium), which continues to push the projectile after it has lost contact with the original agent. But this will not do: when e.g. a javelin-thrower ceases his follow-through, there is nothing pushing the air, and the air should stop moving, and hence the javelin, just as soon as the arm motion is ended. But of course the javelin continues on its course.

267ᵃ2–15

Aristotle's solution to the problem is that the mover does not impart merely motion to the medium, but a power to move. In terms of Aristotle's distinction between active and passive powers (*Met.* Δ12, 1019ᵃ15–23), it imparts the active power of being a mover, as well as the passive power of being moved. While the passive power, or condition of being moved, ceases immediately when the mover is no longer in contact with a body or no longer acting on it, the active power of being a mover apparently fades away gradually. Thus we can understand the gradual ceasing of motion as the effect of a gradual lessening of the power to cause motion in intermediary bodies between the original mover and the projectile.

As an illustration, imagine javelin-thrower A, successive portions of air B, C, and D, and javelin E. According to Aristotle's account, when A ceases his motion (say, his follow-through), at

173

time t_1, B will immediately cease to be in motion, but it will have the power of moving C; so at time t_1, A and B have ceased to be in motion, while C, D, and E are in motion. At t_2, B has spent its motive power, and C ceases to be in motion but retains a power of moving D; at t_3, C has spent its motive power, and D ceases to be in motion but retains a power of moving E; finally, at t_4, D has spent its motive force, and E ceases to move (forward, though it will fall down with natural motion). Aristotle's model provides for a gradual cessation of motion on the part of E.

This seems to be the only place where Aristotle gives the present solution to the problem. One would wish for an argument to support the claim that the intermediary bodies have a power of causing motion, or at least a further analysis of derivative efficient causes. As it is, the present explanation remains rather *ad hoc*: nothing else will explain the phenomenon of projectile motion, so this must be the correct explanation. But an objector might fairly ask why it is that a derivative mover is a mover. Is it not because it is in contact with the original mover? What other event accounts for its being a mover, given that when the original mover (or some original mover) is not in contact with it, it does not move anything? Moreover, since the typical medium of motion is a 'simple' body such as air, it is difficult to see what sort of structural complexity might support the claim that it has some sort of quasi-independent agency to be a mover. The only properties which air seems to have are hotness and wetness (*GC* II. 5, 330ᵇ4), neither of which is an obvious candidate for sustaining an active power of motion. In his explanation of animal generation, Aristotle appeals to motion to account for conception (*GA* I. 22, 730ᵃ19–22), by analogy with a craftsman working on material (much remains unexplained in the biological account, but at least semen is at a higher level of complexity than the elements in Aristotelian theory, so there is room for some latent powers in semen). The heat inherent in air could possibly support a similar account, except that Aristotle seems to have blocked such an account by saying at ᵃ5–6 that the (contiguous) medium stops moving at the same time as the original mover does. But how can any physical medium be a mover without itself being in motion? And why does Aristotle need to rule out motion in the medium anyway? Air around an arrow is in motion—not for reasons which Aristotle would recognize, but at least sense experience does not require that we ascribe perfect rest to the medium.

Presumably what Aristotle wants is a sequential desisting from motion in successive parts of the medium, so that the projectile can lose speed slowly instead of suddenly. But why not have the parts of the medium lose their own motion slowly, and with it their power to cause motion? Perhaps because then we must face the fact that each part of the medium has a kind of internal momentum, which is unexplained in Aristotelian physics.

Given that Aristotle assigns to the medium the potential to cause motion without being in motion, perhaps he could claim that the manifest motion is replaced by some kind of latent motion. But he does not say that, so the mechanism for conserving and transferring motion remains mysterious. One is tempted to think of mass as a means by which motion could be preserved; but the Newtonian notion of mass is foreign to Aristotle, as is the concept of momentum. In general, Aristotle's claim that the medium conserves motion needs further clarification and support. Simplicius cites Alexander's suggestion (1347. 3 ff.) that the medium becomes self-moving in a sense—but not in the sense in which the medium could be a complex of an active and a passive part, as is an animal. Yet then we seem to have a distinction without a difference. In the end, Simplicius is reduced to wishing for a more plausible account (1350. 7–9).

The present problem is an especially interesting one, because it reveals a point of conflict between Aristotelian theory and empirical observation. On Aristotle's theory unnatural or non-natural motion requires the application of force. Since Aristotle recognizes no action at a distance, the force must be applied immediately to the moving body, if not by the mover itself, then by some medium or intermediary. But the medium itself must be impelled to move. On this theory we should anticipate that when the original mover ceases to operate, the whole chain of impulses should cease, and the projectile should fall straight down, describing its natural motion. In response, Aristotle seems to move in the direction of ascribing to the medium an 'impulse' that the first mover imparts (*to endosimon*: Alexander *ap.* Simpl. 1347. 4, 14, 34, 35). Aristotle seems to be flirting with an impetus theory. If so, his suggestions would undermine his own mechanics, and ultimately require a rethinking of his first principles: if an impetus can be impressed on a moving body, why could not an adaptable medium—e.g. the fifth element—preserve unnatural motion indefinitely? To be sure,

more than the present passage would be necessary to overthrow
Aristotle's physical theory. But here, for once, empirical considerations seem to force him to a more sober assessment than usual of
the limits of his theory.

267ᵃ15–20

Aristotle rejects the alleged explanation of 'antiperistasis',
recirculation or 'mutual replacement' (Hardie and Gaye; not
'elasticity' as in Wicksteed's note). As Simplicius explains,
'Antiperistasis occurs when, as one body is being displaced by
another body, there is an exchange of places such that the displacing body takes the place of the displaced body, the displaced body
replaces the next and that the succeeding body, if there are several
bodies, until the last comes to occupy the place of the first displacing body' (1350. 31–6). Plato proposes as an explanation of continued projectile motion (and other phenomena) the cyclical motion
of the medium (*Tim.* 79a–80c). On this view (not developed in
detail by Plato, who concentrates on respiration, making only a
programmatic reference to projectiles, 80a1–2), warm air breathed
out through the mouth forces cool air into the body through pores
in the skin; when it is heated, it escapes back out through the pores,
forcing cool air back into the mouth in inhalation.

But Aristotle objects that recirculation does not solve the
problem: as soon as the air in the original place of the thrower is
replaced, the projectile motion should stop. As Simplicius puts it,
recirculation is a fact, but not an explanation; for instance, walking
will cause recirculation, but recirculation does not cause walking
(1351. 12–16). Recirculation is consistent with the sudden cessation
of projectile motion; a further explanation is still needed of the
observed inertia of the projectile. For John Philoponus's perceptive
criticisms of recirculation, see his *In phys.* 639. 3 ff.

267ᵃ21

'Since in the realm of beings there must be continuous motion': the
continuity of motion is argued in Chs. 1–2, and its cause (continuous circular motion) is explicated in Chs. 7–9. With the present
argument compare also 6. 259ᵃ13–20.

267^{a}25–^{b}6

This section recapitulates the argument of Chs. 5–6.

267^{b}6–7

'The mover, then, must be either at the centre or on the circumference of a circle, for these are the principles of a circle': Aristotle
omits to say that the only kind of motion that is absolutely continuous is circular motion, the point made in Ch. 9. On the centre as
principle of the circle, cf. 9, 265^{b}3. He has not said that the circumference is a principle, though it is natural to think of the centre and
the circumference as the two defining terms of the figure.

267^{b}7–9

267^{b}7–8: 'But things nearest the mover move most swiftly': this
premiss is valid for efficient causes in physical contact with their
objects. But it is not at all clear that it works for all kinds of causes,
and in particular, for final causes. Why should Dulcinea not move
Don Quixote just as powerfully (perhaps more so) for being distant? Even in the case of efficient causes in general, it is not obvious
that the principle holds true. For the master builder is in one
important sense the efficient cause of a building, but he need not be
near the building to exercise control; and certainly he can remain
more distant from the building (except for occasional inspection
tours) than the labourers themselves (cf. *Pol.* VII. 3, 1325^{b}16–23, on
the directing power of the master builder's plans). In the case of a
wheeled vehicle, the ox which provides the motive power may pull
at some distance from the most rapidly moving part of the vehicle,
the rim of the wheel.

Indeed, if we were to put the unmoved mover on the circumference of the cosmos, where precisely would we put it, and why? In
keeping with the reasoning of the present passage, one would
suppose that it must be on the celestial equator (Plato's Circle of
the Same), which is moving most rapidly relative to the earth.
Alexander puzzled over the problem: if the unmoved mover were
at one of the celestial poles, it would not be in motion, but neither
would it be near the most rapidly rotating part of the heavens; if, on

the other hand, it were located at some place of the rotating part of
the heavens, the affected part would be near the cause, but the
unmoved mover would also be moved incidentally, for it would be
rotating with the heavens (Simpl. 1354. 12 ff.). He resolved the
problem by saying that the unmoved mover was 'in the whole
circumference of the outermost sphere' (ibid. 22). Simplicius complains that Alexander had previously seemed to reject the notion
of a physical location for the unmoved mover, on the grounds
that it was incorporeal (lines 26 ff.)—a view which the Neoplatonist
Simplicius favours, and which results in the paradoxical consequence that the unmoved mover is 'everywhere and nowhere'
(1354. 39–1355. 3). Meanwhile, Simplicius gives a report on
Eudemus, whose views he seems to know partly at second hand
from Alexander, partly at first hand. According to Alexander,
Eudemus holds that the unmoved mover is 'in the greatest circle'
which is 'through the poles' (1355. 33), and, as Simplicius himself
reads in Eudemus, 'In a sphere the place round the poles exhibits
the swiftest motion' (lines 35–6)—this after Simplicius has quoted
another passage which he takes as suggesting that there is no physical contact between the unmoved mover and the heavenly sphere
(lines 28 ff.—although the passage quoted is in fact ambiguous).
The remark reported at lines 35–6 makes sense if it is interpreted as
meaning a great circle equidistant from the poles—i.e. the celestial
equator. The equator is not 'through the poles' (line 33), of course,
but it does bound a plane bisecting the diameter drawn between
the poles.

From the commentators we can infer at least (i) that there was
no authoritative interpretation of Aristotle's doctrine in the
Peripatetic tradition; and (ii) that the understanding of the unmoved mover as an incorporeal being without a unique location
emerged only slowly and with difficulty. Aristotle, for his part, does
not seem to have worried about the danger of assigning to his
unmoved mover a place on a moving body, or about the further
paradoxes of assigning it any location at all (on which see below on
267^b19–26).

267^b9–17

One of the triumphs of modern engineering has been the translation of separate motions into continuous motion—e.g. the back-

and-forth motion of a piston in a steam-engine or internal combustion engine to circular motion by means of a flywheel. But, considered in its totality, the functioning of the engine is not, in fact, completely continuous. The rejection of physical sources of motion seems to eliminate the possibility that the unmoved mover pushes or pulls the outermost sphere of the cosmos. But how does it cause motion? Curiously, Aristotle does not say anywhere in this treatise. Is it because the explanation of how the first unmoved mover operates goes beyond physics to first philosophy, i.e. metaphysics? Of course, Aristotle does supply an answer to the question in *Met.* Λ7, where he makes the unmoved mover a final cause of motion. But why could he not say that much here, even if he cannot go on to fill in the account by discussing the divine life that the unmoved mover enjoys? The scheme of the four causes is very much a part of physics proper, having been expounded in *Phys.* II. 3 and 7 and applied innumerable times throughout the *Physics* and other works of natural philosophy.

Simplicius argues that the first unmoved mover is a cause not only in the sense of being a final cause—which everyone in his day, as in ours, would accept—but also in the sense of being an efficient cause (1360. 24 ff.), and his master Ammonius wrote a whole book defending the thesis (ibid. 1363. 8–10). Simplicius's arguments include citations of Plato's views in the *Timaeus*—evidence not relevant to the debate unless one happens to believe in the essential harmony of Plato and Aristotle—and inferences from approving remarks which Aristotle makes about the role of *Nous* in Anaxagoras, which require a good deal of reading between the lines. But he does point out rightly that the unmoved mover fits the definition of an efficient cause—'whence the first source of change or rest' (*Phys.* II. 3, 194^{b}29–30; Simpl. 1361. 12 ff.). The examples which Aristotle adduces do not obviously suggest an application to the first unmoved mover, and it is at least possible that Aristotle originated his fourfold distinction without reference to such an entity. But the real question is whether, given his definition of the efficient cause, it includes the unmoved mover willy-nilly. One curious fact remains: that Aristotle never acknowledges the alleged fact that the unmoved mover is an efficient cause (a problem of which Simplicius is well aware: 1363. 12–14). We are back to the problem suggested in the Concluding Aporia to Ch. 5. Recently, Judson (1994) has argued for the view that the unmoved mover is indeed an efficient cause. This may be so, but if it is, *Physics* VIII itself does not supply adequate evidence for the view.

267ᵇ19–26

Aristotle sets up an elegant dilemma to show that the unmoved mover is without size. Infinite size is ruled out by Aristotelian physics; finite size is ruled out by arguments set up at the beginning of the chapter. Hence the unmoved mover has no size. But what, then, is it, and how does it act? This problem seems to be acute, because of other things Aristotle says or assumes. The unmoved mover has a location, specifically at or on the outermost heavenly sphere (see above on ᵇ7–9). But why should proximity be a consideration? It cannot cause motion by physical means, or it would have to act by contact. Furthermore, how can it be in physical contact when it has no dimensions (cf. Eudemus *ap.* Simpl. 1357. 21–3)? Bodies (or processes) are in contact, by definition, when their extremes are together (*Phys.* V. 3, 226ᵇ23), but the unmoved mover, by virtue of being dimensionless, can have no extremes. Moreover, if it has a place, it must either be on the rotating heavenly sphere or not. If it is on the sphere, it is not absolutely unmoved, but is in incidental motion. If, on the other hand, it is not on the sphere, where is it? At *Cael.* I. 9, 279ᵃ17–22 Aristotle suggests an answer in a lyrical passage: 'It is clear then that there is neither place nor void nor time outside [the heaven]. Accordingly, the things there (*ta'kei*(!)) cannot be in place, nor does time age them, nor is there any change in the things ranged beyond the outermost orbit; but unaltered and impassible they continue through their whole existence living the best and most self-sufficient of lives.' We would expect the consequence of the first sentence to be that there is nothing outside the heaven; but instead we discover that there are things there which live the best kind of life untrammelled by the ills born of space and time. After Aristotle's hard-headed views about nature, this seems like pure fantasy, embodying the most striking paradoxes. But what precisely *are* the things out there? Aristotle gives no clear answer. His use of the plural makes us think that he is including more things than the first unmoved mover—unless he is invoking the royal plural.

On the other hand, what does it mean for something without parts or size to have a location for Aristotle? His account of place at *Phys.* IV. 1–5 presupposes that things which occupy places are physical bodies. On his mathematical theory an indivisible (*adiaireton*, the same word used in ^{b}25) quantity, if it has no location, is a unit, while if it has a location, it is a point (*Met. Δ*6, 1016^{b}24–6, 29–31). It seems, then, that the unmoved mover would be at least equivalent to a mathematical point. But, according to Aristotle, no point can be a substance (*Met.* N3, 1090^{b}5–13).

If, on the other hand, the unmoved mover does not cause motion by physical means and does not have parts or size, why should it have a location at all? We seem to have a situation like the one Descartes gets himself into when he asks where the soul (a non-extended thinking thing) interacts with the body (an extended, non-thinking thing). He suggests that the pineal gland is the locus of interaction because of its proximity to the brain (and lack of any known function). But to make sense of this, we must suppose that the soul, a thinking, unextended thing, either has a location or is capable of having a location. By parity of reasoning, we may ask what it means for the unmoved mover to have a spatial location. At most, it can be like a geometrical point with a position (though not, in Aristotle's sense, a place). But why does it need a position? Not to have contact with the heavens, since it could not have physical contact; nor could its physical contact, if it had it, be responsible for its infinite power to cause motion. Is Aristotle looking for a pineal gland?

It has often been suggested that the unmoved mover is a relic of the Platonic Forms, a transcendent source of causation for the physical world. The advantage enjoyed by Plato is that he does not try to locate the Forms in the world of becoming. That Aristotle does so try suggests that he is struggling with the distinction between physical and non-physical existents. Unlike Plato, Aristotle is unwilling to posit the existence of anything outside the physical cosmos. But his unmoved mover has properties which make it unfit to exist like other members of the physical cosmos. He seems to be left with a transcendent being having at least one physical property, position, making it an ontological hybrid.

267^{b}20–2:   'That there cannot be an infinite magnitude has already been proved in the *Physics*': at III. 5, 204^{a}34 ff.; also at *Cael.* I. 5–7.

267^{b}22–4:   'That a finite magnitude cannot have infinite power . . . has just now been proved': at 266^{a}12–^{b}27.

APPENDIX I

Outline of the Argument

Introduction: Did motion always exist? ($250^{b}11$)

(I) Motion has always existed in the world (conclusion drawn at $252^{b}5$).
  - (A) There are two ways in which it might not always have existed ($250^{b}23$).
    - (1) Motion might begin after an infinite period of rest.
    - (2) Motion and rest might alternate.
  - (B) Intermittent motion is impossible ($251^{a}8$).
    - (1) Motion is the actuality of the movable *qua* movable.
    - (2) Thus, there must be movers and movables.
    - (3) These must either (i) be everlasting or (ii) come into existence.
    - (4) (3 ii) is impossible, for it presupposes a prior change: namely, the coming into being of the movers and movables ($251^{a}17$).
    - (5) (3 i) is impossible.
      - (a) Irrational powers interact automatically ($251^{a}28$).
      - (b) If they did not interact, some change prior to the first change must have taken place to cause them to interact.
      - (c) (Motion and time are coextensive; there could be no time if there were no motion ($251^{b}10$)).
    - (6) If motion ceases, some change posterior to the last change must take place to end the interaction of powers ($251^{b}28$).
  - (C) It is inadequate to say that intermittent motion is natural ($252^{a}5$).
    - (1) Empedocles takes intermittent motion as natural.
    - (2) But it is not natural.
      - (a) Nature is a cause of order.
      - (b) There will be no ratio between times of rest and of motion.
      - (c) If there is no ratio, there is no order.
      - (d) But intermittent motion would be more natural than a single beginning of motion.
    - (3) Empedocles needs to argue for his position ($252^{a}22$).

183

(II) Objections to (I) can be met.
    (A) There are several objections to (I) (252^b7).
        (1) Every change is between contraries, which limit motion.
        (2) We see from experience of inanimate objects that movable things do not always move.
        (3) We see from experience of animate objects that movable things do not always move.
            (a) Animate objects initiate their own motion.
            (b) If this can happen in the microcosm, then it can happen in the macrocosm.
            (c) If it can happen in the cosmos, then it can happen outside it.
    (B) Objection (A1) can be met (252^b28).
        (1) Changes between contraries are indeed limited.
        (2) But it is possible that some changes are not between contraries.
    (C) Objection (A2) can be met (253^a2).
    (D) Objection (A3) can be met (253^a7).
        (1) An animal is not responsible for initiating motion in place.
        (2) Changes in the environment can initiate the motion.

(III) Some things are always unmoved, some always moving, and some vary between motion and rest.
    (A) There are three possible alternatives with respect to cosmic motion.
        (1) All things are always at rest.
        (2) All things are always in motion.
        (3) Some things are in motion, some at rest.
            (a) All movables are always in motion; all things at rest are always at rest.
            (b) All things vary between motion and rest.
            (c) Some things are always in motion, some always at rest, and some things vary between motion and rest.
    (B) (A1) is untenable (253^a32).
    (C) (A2) is untenable (253^b6).
        (1) Increase and decrease are not continuous processes.
        (2) Alteration is not a continuous process.
        (3) Locomotion does not take place in all bodies.
    (D) (A3a) is untenable.
    (E) Recapitulation: (A1), (A2), and (A3a) are untenable (254^a15).

(IV) All things that are in motion are moved by something.
    (A) We must distinguish different senses of motion (254^b7).
        (1) We are concerned with intrinsic, not incidental, motion.

    (2) Intrinsic motions are caused by either the agent itself or by another cause.

    (3) Some intrinsic motions are caused by the thing itself, some by another.

    (4) Some intrinsic motions happen by nature, some by force.

(B) Of motions caused by another, some happen by nature, some contrary to nature (254^b20).

    (1) Bodies moved contrary to nature are clearly moved by another.

    (2) Animals etc. are moved by something.

(C) In the case of natural motion of the elements, it is problematic how it is caused by another (254^b33).

    (1) The elements lack the intentional dimension which accounts for animal motion.

    (2) They lack the complexity presupposed by self-motion.

    (3) But they too are moved by something.

(D) We must distinguish the causal factors at work in elemental motion (255^a20).

    (1) Movers can move either by nature or contrary to nature.

    (2) There are different senses of 'potency' (255^a30).

        (a) In one sense, X is potentially F if it can acquire the property F.

        (b) In another sense, X is potentially F if it is F but is not exercising the property.

        (c) The heavy is potentially light in the first sense, whereas the light which is in the lower regions is potentially light in the second sense.

(E) The causal factors show how elemental motion is caused by another (255^b13).

    (1) Elements are potentially light in the aforesaid ways.

    (2) An agent which removes an obstacle activates the higher potency.

    (3) The elements have a passive power of being acted on.

    (4) Thus, all bodies in motion are moved by something.

(V) A first mover is unmoved.

    (A) If the first mover is moved, it is moved by itself (256^a4).

        (1) Otherwise there will be an infinite series of movers.

        (2) Or there will be an infinite series of intermediaries (256^a21).

        (3) This can be shown by making a distinction (256^b3).

            (a) If everything moved is moved by a moved, this fact obtains either incidentally or intrinsically.

            (b) If incidentally, an impossibility results.

        (c) If intrinsically, absurdities result (256^b27).
           (i) If the mover is moved with the same kind of motion, an impossibility results.
           (ii) If the mover is moved with a different kind of motion, absurdities result.

(B) A self-mover moves by virtue of a part being the mover, another part the moved (257^a27).
  (1) The self-mover cannot move and be moved with the same motion.
  (2) The self-mover cannot move itself as a whole (257^a33).
  (3) The self-mover cannot move itself by having the parts move each other (257^b13).
    (a) If two parts move each other, we cannot identify a first mover.
    (b) If one part moves the other part, there is no reason for the second to move the first.
    (c) There is no need for the mover to be moved in turn.
    (d) If the parts moved each other, they would be moved with the same motion that they cause.
  (4) The self-mover cannot have one or more parts that move themselves (257^b26).
    (a) If it is moved in virtue of a part that moves itself, that part is the primary self-mover.
    (b) If it is moved in virtue of the whole, the parts move themselves incidentally.
  (5) The self-mover consists of an unmoved and a moved part (258^a3).
    (a) Something that is moved but does not cause motion is not an integral part of a mover.
    (b) The part that moves something by being moved is necessary but not self-sufficient as a mover.
    (c) There is a problem about whether we can take anything away from a continuous self-mover; this problem can be met (258^a27).
  (6) Conclusion: the original mover of any motion is unmoved.

(VI) The first mover of the cosmos is everlasting and unmoved.
  (A) Even if some unmoved movers perish, not all of them can perish (258^a16).
    (1) Motion is everlasting and continuous.
    (2) Perishable things are not everlasting and continuous.
    (3) Thus perishable things cannot explain cosmic motion.
  (B) There is at least one everlasting mover, and one is sufficient (259^a6).

(C) Everlasting motion must be continuous and have a single mover (259^{a}13).
(D) The previous argument implies this (259^{a}20).
    (1) There are things which vary between rest and motion, and things which are always in motion, and things which are always unmoved.
    (2) Animate objects seem to move themselves (259^{b}1).
    (3) Factors outside the animal cause some motions.
    (4) The first mover is moved incidentally.
    (5) Hence this is not the cause of continuous motion (259^{b}20).
(E) The first moved body will also be everlasting (260^{a}11).

(VII) Locomotion is the primary kind of motion.
  (A) Growth presupposes locomotion (260^{a}20).
    (1) There are three kinds of motion.
    (2) One of these, growth, requires alteration (assimilation).
    (3) Alteration requires locomotion.
  (B) Even on the popular account of change, locomotion is presupposed (260^{b}7).
    (1) Affections are caused by condensation and rarefaction.
    (2) Condensation and rarefaction are caused by aggregation and segregation.
    (3) Aggregation and segregation are instances of locomotion.
  (C) Locomotion is primary in the order of dependence (260^{b}15).
    (1) There are several senses of 'primary'.
    (2) One of these is what other things depend on for existence.
    (3) Only locomotion can be continuous.
    (4) Other motions depend on locomotion, while it does not depend on them.
  (D) Locomotion is primary in order of generation (260^{b}29).
    (1) Objection: locomotion comes after generation and motions subsequent to it in an animal.
    (2) Reply: there must be a prior mover moving in place to generate the animal.
    (3) Generation belongs only to perishable things.
  (E) Locomotion is primary in nature (261^{a}13).
    (1) What is posterior in generation is prior in nature.
    (2) Locomotion is last in generation.
    (3) The subject of locomotion does not change its nature.
    (4) The self-mover moves itself in place.
  (F) Only locomotion is continuous (261^{a}27).
    (1) All other changes are from contrary to contrary or contradictory to contradictory.
    (2) Changes to a contrary must be preceded by rest.

(3) Changes to contradictories must be preceded by a time interval.

(4) Motion is opposed to the contrary motion and to rest (261^{b}15).

(VIII) There can be some everlasting motion which is one and continuous: namely, circular motion.

(A) Motions other than circular motion cannot be continuous (261^{b}28).

(1) Motions are straight, circular, or a combination.

(2) If one element of combined motion is not continuous, neither is combined motion.

(3) Since straight lines are finite, motion along them must double back in a contrary direction.

(4) Contrary motions are not continuous.

(B) Motion in a straight line is not continuous (262^{a}12).

(1) Even motion on a circle which doubles back is not continuous.

(2) Continuous motion cannot arrive at and depart from a middle point.

(3) This analysis solves theoretical problems about motion (262^{b}8).

(a) Two bodies setting out at the same time at the same speed to cover the same distance must arrive together.

(b) One cannot stop at a mid-point.

(4) A moving body that changes direction must pause at the turning-point (262^{b}21).

(5) Zeno's stadium is a problem about crossing mid-points (263^{a}4).

(6) The problem is solved by treating the mid-points as potential divisions (263^{a}11).

(7) We must take a point of time marking a change as belonging to the later time to avoid contradictions (263^{b}9).

(C) General theoretical arguments also support this position (264^{a}7).

(1) If motion that doubles back is continuous, the moving body will be pursuing contrary motions at the same time.

(2) What moves with intermittent motion must previously have been at rest in the opposite state (264^{a}21).

(3) If motion is continuous, incompatible states will coexist (264^{b}1).

(4) The continuity of time does not imply the continuity of motion (264^{b}6).

(D) Motion along a circle is one and continuous (264^{b}9).
    (1) Circular motion does not retrace the same path, but motion on a straight line does.
    (2) Thus a body moving along a straight line pursues opposite motions at the same time.
(E) No other motions are continuous (264^{b}28).
    (1) Other kinds of change traverse the same stretches.
    (2) Natural philosophers are wrong to say there is continuous flux.

(IX) Circular motion is primary.
    (A) It is simple and complete (265^{a}16).
        (1) If motion along a straight line doubles back, it is not simple.
        (2) If it does not double back, it is not complete.
    (B) It can be everlasting, while other motions must come to a halt (265^{a}24).
    (C) Only circular motion can be one and continuous (265^{a}27).
        (1) There is a determinate beginning, middle, and end of a straight line.
        (2) A moving body must rest at the beginning and end.
        (3) There is no beginning, middle, and end on a circle.
        (4) The centre constitutes the beginning, middle, and end.
        (5) Hence a body moving in a circle need not rest.
    (D) The circle is primary as a measure (265^{b}8).
        (1) Whatever measures other motions is primary, and conversely.
        (2) The circle measures other motions.
    (E) Only circular motion can be uniform (265^{b}11).
    (F) All philosophers agree that locomotion is primary (265^{b}16).
        (1) Aggregation and segregation presuppose locomotion.
        (2) Condensation and rarefaction presuppose locomotion.
        (3) Those who make soul the original cause of motion make locomotion primary.
        (4) Locomotion is the primary sense of motion.
    (G) Conclusion: cosmic motion is everlasting circular locomotion, caused by an unmoved mover (266^{a}6).

(X) The first unmoved mover is indivisible and without parts and magnitude.
    (A) An infinite motion cannot be caused by a finite mover (266^{a}12).
    (B) An infinite power cannot belong to a finite body (266^{a}24).
    (C) A finite power cannot exist in an infinite magnitude (266^{b}6).

(D) Projectiles are moved in part by the medium (266^b27).
  (1) Why does a projectile not stop moving when its thrower stops its motion?
  (2) The mover imparts a power of being a mover to the medium (267^a2).
  (3) 'Recirculation' does not solve the problem.
(E) The first mover must be unmoved (267^a21).
  (1) Since cosmic motion is continuous, it must originate with the single motion of a single body moved by a single mover.
  (2) The mover either must be unmoved or must depend on an unmoved mover.
  (3) It will cause motion without doing work.
  (4) It must be located at the circumference of the cosmic sphere.
(F) A mover that pushes or pulls (does physical work) cannot cause continuous motion (267^b9).
(G) The unmoved mover is indivisible and without parts and magnitude (267^b17).

APPENDIX II

Two Senses of *Kinēsis*

There is a technical problem for Aristotle implicit in the discussion at 8, 262^{b}10 ff (discussed by Waterlow 1982: 145–6; White 1992: 104 ff.): namely, how it is possible for a motion to take place if it consists of motions. According to *Phys.* VI. 5, any body in motion has already been in motion. According to *Phys.* V. 4, a motion is bounded by periods of rest. But it also appears from VI. 5 and other passages that Aristotle's completed motions are constituted by what White (p. 102) calls 'sub-motions', which, because they are defined arbitrarily by segments of the whole motion which they comprise, are infinite in number. What is problematic is that these sub-motions are not bounded by periods of rest, yet somehow they too are motions. In response to this problem, White thinks that Aristotle refines or corrects (p. 106) his account in Book VI in the present passage: the intermediate point of a larger motion is a goal only potentially, not actually.

It seems to me that the whole problem is a bogus one, to which no correction in Aristotelian theory is necessary. There is (at least) one fundamental ambiguity in the term *kinēsis* ('motion') which continues to bedevil good scholars. The motion whose defining conditions are given in Book V. 4 is completed motion, in ontological terms an event, complete, as Aristotle says of plot-events in the *Poetics* (7, 1450^{b}23–7), with a beginning, a middle, and an end. By contrast, the motion that is defined in *Phys.* III. 1 as the 'realization of what is potentially, as such' (201^{a}10–11) is an activity of moving, which, as soon as it reaches its end-point, is no more. In ontological terms, it is a process. Now a process can be part of an event, but it is of itself incomplete, a segment of an event; it is expressed by a continuous tense—e.g. 'I am walking to the market'—whereas an event is expressed by a tense embodying the 'perfective' aspect—e.g. 'I walked to the market'. Processes are *essentially* incomplete or indeterminate, whereas events are essentially complete, or at least completable. They are, or can be, as the linguistic categories aptly suggest, different aspects of the same situation. This verbal distinction between aspects corresponds to a nominal distinction between count-nouns (there was *a* motion, there were two motions) and mass-nouns (there was *some* motion), as Mourelatos (1978) has shown. We can count episodes of complete motion, but not ongoing processes.

Motion processes—e.g. stretches of walking—go to make up motion events—e.g. walking from home to the market-place. But the two types of motion belong to different ontological categories, and, according to Aristotle, the event category is prior: walking processes can be defined only relative to some walking event. This distinction may indeed cause problems for Aristotle's notion of continuous everlasting circular motion, but it is, at a basic ontological level, perfectly coherent, even if it could be disputed in point of fact. (For more on this subject see Mourelatos 1978, Graham 1980, also Penner 1970.)

Aristotle never clearly disambiguates his two major senses of motion. But they exist side by side, and will be irreconcilable if we do not disambiguate them ourselves. If we do, the alleged conflict between accounts of motion in the *Physics* disappears: it is a truism that motion events are composed of motion processes. Motion processes by their very nature are indeterminate, and must be individuated arbitrarily; motion events are determinate, and are individuated by their natural termini. Aristotle could say in this context, as he does in others, that component processes are only potentially, not actually, infinite in number.

The relationship between motion processes and motion events is analogous to that between mass-nouns and count-nouns: the latter are intrinsically determinate and countable, whereas the former can be individuated and counted only by means of some arbitrary measure: a *cup* of water, a *pinch* of salt, a *bowl* of soup, a dozen *head* of cattle. Similarly, compare 'a bit of walking' and 'a walk'. Aristotle holds that motion events are comprised of motion processes, but that the latter in no way constitute the former—i.e. the former are not reducible to, or supervenient on, the latter. The case is wholly parallel to the relationship between concrete substances and matter: although the substance is comprised of matter, matter does not constitute or define the substance. And of course, linguistically concrete substances are expressed by count-nouns, matter by mass-nouns.

In *Met.* Θ7, 1049ᵃ18 ff. Aristotle notes that what is potentially something—i.e. the matter that is potentially a substance—is referred to by expressions of the form 'thaten' (*ekeininon*). In expounding his distinction, he notes that the subject is a This (*tode ti*); i.e. it can be singled out by ostension, whereas an incidental attribute is not a This. And he includes in his examples the case of the subject which 'is not a walk or a motion but walking or moving, i.e. thaten' (ᵃ33). This analysis seems to be Aristotle's recognition of the link between mass-nouns and process predications of verbs. As substance and matter are ontologically and (often) linguistically distinct but related, so are events and processes—different but compatible aspects, we might say, of the same situation. Thus there is no inherent conflict between Aristotle's definition of motion (as process) in *Phys.* III. 1 and his determination of the identity conditions of motion (as event) in *Phys.* V. 4; there is, however, a serious need for clarification, one which Aristotle himself never seems to address.

SELECT BIBLIOGRAPHY

Names and abbreviations marked by an asterisk (*) are referred to by that name or abbreviation alone in the commentary.

Abbreviations of journals and collections

AGP   *Archiv für Geschichte der Philosophie*
AJP   *American Journal of Philology*
AP   *Ancient Philosophy*
CIAG   *Commentaria in Aristotelem Graeca*, 23 vols. (Prussian Academy, Berlin, 1882–1909), with Supplementum Aristotelicum, 3 vols. (1882–1903)
CP   *Classical Philology*
CQ   *Classical Quarterly*
OSAP   *Oxford Studies in Ancient Philosophy*
PAS   *Proceedings of the Aristotelian Society*
PBACAP   *Proceedings of the Boston Area Colloquium in Ancient Philosophy*
PQ   *Philosophical Quarterly*
PR   *Philosophical Review*

ALLAN, D. J. (1954), 'The Problem of Cratylus', *AJP* 75: 271–87.
*AQUINAS, ST THOMAS, *In VIII Libros Physicorum Aristotelis* (Rome, apud sedem Commisionis Leoninae, 1884). Translation: *Commentary on Aristotle's Physics*, trans. R. J. Blackwell, R. J. Spath, and W. E. Thirlkel (New Haven: Yale University Press, 1963).
ARCHER-HIND, R. D. (1888), *The Timaeus of Plato* (London: Macmillan and Co.).
BALLEW, LYNNE (1979), *The Straight and the Circular* (Assen: Van Gorcum).
BARNES, JONATHAN (1979a), 'Parmenides and the Eleatic One', *AGP* 61: 1–21.
——(1979b), *The Presocratic Philosophers*, 2 vols. (London: Routledge & Kegan Paul).
BODNÁR, ISTVÁN M. (1997), 'Movers and Elemental Motions in Aristotle', *OSAP* 15: 81–117.
*BONITZ, HERMANN (1870/1961), *Index Aristotelicus* (Berlin: W. de Gruyter).

BOSTOCK, DAVID (1991), 'Aristotle on Continuity in *Physics* VI', in Judson (1991), 179–212.

BURNET, JOHN (1892/1930), *Early Greek Philosophy*, 4th edn. (London: Adam & Charles Black).

CHERNISS, HAROLD (1935), *Aristotle's Criticism of Presocratic Philosophy* (Baltimore: Johns Hopkins University Press).

——(1944), *Aristotle's Criticism of Plato and the Academy*, i (Baltimore: Johns Hopkins University Press).

COHEN, SHELDON M. (1994), 'Aristotle on Elemental Motion', *Phronesis*, 39: 150–9.

*CORNFORD, F. M. See Wicksteed and Cornford (1929, 1934).

CORNFORD, F. M. (1937), *Plato's Cosmology* (London: Routledge & Kegan Paul).

——(1939), *Plato and Parmenides* (London: Routledge & Kegan Paul).

DEVEREUX, DANIEL (1988), 'The Relationship between Theophrastus' *Metaphysics* and Aristotle's *Metaphysics* Λ', in Fortenbaugh and Sharples (1988), 167–88.

DIELS, HERMANN (1879), *Doxographi Graeci* (Berlin: W. de Gruyter).

*DK: Diels, Hermann, and Kranz, Walther (1951), *Die Fragmente der Vorsokratiker*, 6th edn., 3 vols. (Dublin: Weidmann).

EASTERLING, H. J. (1976), 'The Unmoved Mover in Early Aristotle', *Phronesis*, 21: 252–65.

FORTENBAUGH, W. W., and SHARPLES, R. W. (1988) (eds.), *Theophrastean Studies* (New Brunswick, NJ: Transaction Books).

FREDE, DOROTHEA (1971), 'Theophrasts Kritik am unbewegten Beweger des Aristoteles', *Phronesis*, 16: 65–79.

FURLEY, DAVID (1967), *Two Studies in the Greek Atomists* (Princeton: Princeton University Press).

GILL, MARY LOUISE (1989), *Aristotle on Substance: The Paradox of Unity* (Princeton: Princeton University Press).

——and LENNOX, JAMES G. (1994) (eds.), *Self-Motion: From Aristotle to Newton* (Princeton: Princeton University Press).

GRAHAM, DANIEL W. (1980), 'States and Performances: Aristotle's Test', *PQ* 30: 117–30.

——(1987), *Aristotle's Two Systems* (Oxford: Clarendon Press).

——(1988), 'Symmetry in the Empedoclean Cycle', *CQ* 38: 297–312.

——(1989), 'Aristotle's Definition of Motion', *AP* 8: 209–15.

——(1991), 'Socrates, the Craft Analogy, and Science', *Apeiron*, 24: 1–24.

——(1996), 'The Metaphysics of Motion: Natural Motion in *Physics* II and *Physics* VIII', in W. Wians (ed.), *Aristotle's Philosophical Development* (Lanham, Md.: Rowman and Littlefield), 171–92.

——(1997), 'Heraclitus' Criticism of Ionian Philosophy', *OSAP* 15: 1–50.

GUTHRIE, W. K. C. (1933–4), 'The Development of Aristotle's Theology', *CQ* 27: 162–71; 28: 90–8.

——(1939) (trans.), *Aristotle: On the Heavens*, Loeb Classical Library (Cambridge, Mass.: Harvard University Press).

——(1962), *A History of Greek Philosophy*, i (Cambridge: Cambridge University Press).

——(1965), *A History of Greek Philosophy*, ii (Cambridge: Cambridge University Press).

HACKFORTH, R. (1959), 'Plato's Cosmogony (*Timaeus* 27D ff.)', *CQ*, NS 9: 17–22.

*HARDIE, R. P., and GAYE, R. K. (1930) (trans.), *Aristotle: Physics* (Oxford: Clarendon Press).

HEIDEL, W. A. (1906), 'Qualitative Change in Pre-Socratic Philosophy', *AGP* 19: 333–79.

HINTIKKA, JAAKKO (1973), *Time and Necessity* (Oxford: Clarendon Press).

HOCUTT, M. (1974), 'Aristotle's Four Becauses', *Philosophy*, 49: 385–99.

HÖLSCHER, UVO (1965), 'Weltzeiten und Lebenskyklus', *Hermes*, 93: 7–33.

HUSSEY, EDWARD (1983), *Aristotle's Physics Books III and IV* (Oxford: Clarendon Press).

INWOOD, BRAD (1992), *The Poem of Empedocles* (Toronto: University of Toronto Press).

IRWIN, TERENCE (1988), *Aristotle's First Principles* (Oxford: Clarendon Press).

JAEGER, WERNER (1923), *Aristoteles: Grundlegung einer Geschichte seiner Entwicklung* (Berlin, Weidmann).

JUDSON, LINDSAY (1983), 'Eternity and Necessity in *De Caelo* I.12', *OSAP* 1: 217–54.

——(1991) (ed.), *Aristotle's Physics: A Collection of Essays* (Oxford: Clarendon Press).

——(1994), 'Heavenly Motion and the Unmoved Mover', in Gill and Lennox (1994), 155–71.

*KRS: Kirk, G. S., Raven, R. E., and SCHOFIELD, M. (1957/1983), *The Presocratic Philosophers*, 2nd edn. (Cambridge: Cambridge University Press).

KAHN, CHARLES (1960), *Anaximander and the Origins of Greek Cosmology* (New York: Columbia University Press).

KERSCHENSTEINER, JULA (1962), *Kosmos: Quellenkritische Untersuchungen zu den Vorsokratikern* (Munich: C. H. Beck).

KIRK, G. S. (1951*a*), 'Natural Change in Heraclitus', *Mind*, 60: 35–42.

——(1951*b*), 'The Problem of Cratylus', *AJP* 72: 225–53.

——(1954), *Heraclitus: The Cosmic Fragments* (Cambridge: Cambridge University Press).

KOSMAN, L. A. (1969), 'Aristotle's Definition of Motion', *Phronesis*, 14: 40–62.

LANG, HELEN S. (1984), 'Why Fire Goes Up: An Elementary Problem in Aristotle's Physics', *Review of Metaphysics*, 38: 69–106.

*LSJ: Liddell, H. J., and Scott, R. (1940), *A Greek–English Lexicon*, 9th edn., rev. H. S. Jones and R. McKenzie (Oxford: Clarendon Press).

LAKS, ANDRÉ, and MOST, GLENN W. (1993), *Théophraste: Métaphysique* (Paris: Les Belles Lettres).

LONG, A. A. (1974), 'Empedocles' Cosmic Cycle in the 'Sixties', in A. P. D. Mourelatos (ed.), *The Pre-Socratics* (Garden City, NY: Doubleday), 397–425.

LOVEJOY, ARTHUR O. (1936), *The Great Chain of Being* (Cambridge, Mass.: Harvard University Press).

LURIA, SOLOMON (1933), 'Die infinitesimallehre der antiken Atomisten', *Quellen und Studien zur Geschichte der Mathematik*, 2: 106 ff.

MARCOVICH, M. (1965), 'Herakleitos', in Georg Wissowa *et al.* (eds.), *Paulys Encyclopädie der classischen Altertumswissenschaft* (Stuttgart: Alfred Druckenmüller Verlag), suppl. vol. x, cols. 246–320.

MATSON, WALLACE I. (1988), 'The Zeno of Plato and Tannery Vindicated', *La Parola del passato*, 43: 312–36.

MOHR, R. D. (1985), *The Platonic Cosmology* (Leiden: E. J. Brill).

MORAVCSIK, J. M. E. (1974), 'Aristotle on Adequate Explanations', *Synthese*, 28: 3–17.

——(1991), 'What Makes Reality Intelligible? Reflections on Aristotle's Theory of *Aitia*', in Judson (1991), 31–47.

MOST, GLENN W. (1988), 'The Relative Date of Theophrastus' *Metaphysics*', in Fortenbaugh and Sharples (1988), 224–33.

MOURELATOS, A. P. D. (1978), 'Events, Processes, and States', *Linguistics and Philosophy*, 2: 415–34.

NUSSBAUM, MARTHA C. (1982), 'Saving Aristotle's Appearances', in M. Schofield and M. C. Nussbaum (eds.), *Language and Logos* (Cambridge: Cambridge University Press), 267–93.

O'BRIEN, DENIS (1969), *Empedocles' Cosmic Cycle* (Cambridge: Cambridge University Press).

——(1995), 'Empedocles Revisited', *AP* 15: 403–70.

OSBORNE, CATHERINE (1987), 'Empedocles Recycled', *CQ* 37: 24–50.

OWEN, G. E. L. (1961), 'Τιθέναι τὰ φαινόμενα', in S. Mansion (ed.), *Aristote et les problèmes de la méthode* (Paris: Publications universitaires), 83–133.

——(1966), 'Plato and Parmenides on the Timeless Present', *Monist*, 50: 317–40.

PENNER, TERRY (1970), 'Verbs and the Identity of Actions—A Philosophical Exercise in the Interpretation of Aristotle', in O. P. Wood and G. Pitcher (eds.), *Ryle: A Collection of Critical Essays* (Garden City, NY: Doubleday), 393–460.

*PHILOPONUS, IOHANNES (1887–8), *In Aristotelis Physica Commentaria*, ed. H. Vitelli, *CIAG* 14.2.

POPPER, KARL (1959), 'Back to the Presocratics', *PAS*, NS 59: 1–24.

RAVEN, J. E. (1948), *Pythagoreans and Eleatics* (Cambridge: Cambridge University Press).

REINHARDT, KARL (1916), *Parmenides und die Geschichte der griechischen Philosophie* (Bonn: Friedrich Cohen).

ROBINSON, T. M. (1987), 'Understanding the *Timaeus*', *PBACAP* 1: 103–19.

*ROSS, W. D. (1936), *Aristotle's Physics* (Oxford: Clarendon Press).

——(1957), 'The Development of Aristotle's Thought', *Proceedings of the British Academy*, 43: 63–78.

RYLE, GILBERT (1949), *The Concept of Mind* (London: Hutchinson & Company).

SEECK, G. A. (1969), 'Leicht-schwer und die unbewegte Beweger', in I. Düring (ed.), *Naturphilosophie bei Aristoteles und Theophrast* (Heidelberg: Lothar Stiehm Verlag), 210–16.

*SIMPL[ICIUS] (1882, 1885), *In Aristotelis Physica Commentaria*, ed. H. Diels, *CIAG* 9–10.

——(1884), *In Aristotelis De Caelo Commentaria*, ed. J. L. Heiberg, *CIAG* 7.

SOLMSEN, FRIEDRICH (1960), *Aristotle's System of the Physical World* (Ithaca, NY: Cornell University Press).

——(1965), 'Love and Strife in Empedocles' Cosmology', *Phronesis*, 10: 109–48.

SORABJI, RICHARD (1988), *Matter, Space, and Motion* (Ithaca, NY: Cornell University Press).

STOKES, MICHAEL C. (1962–3), 'Hesiodic and Milesian Cosmogonies', *Phronesis*, 7: 1–37; 8: 1–34.

——(1971), *One and Many in Presocratic Philosophy* (Washington: Center for Hellenic Studies).

TANNERY, PAUL (1887/1930), *Pour l'histoire de la science hellène*, 2nd edn. (Paris: Gauthier-Villars et Cie).

TAYLOR, A. E. (1928), *A Commentary on Plato's Timaeus* (Oxford: Clarendon Press).

THEILER, WILLY (1925/1965), *Zur Geschichte der teleologischen Naturbetrachtung bis auf Aristoteles*, 2nd edn. (Berlin: W. de Gruyter).

*THEMISTIUS (1900), *In Aristotelis Physica Paraphrasis*, ed. H. Schenkl, *CIAG* 5.2.

VENDLER, ZENO (1967), *Linguistics in Philosophy* (Ithaca, NY: Cornell University Press).

VLASTOS, GREGORY (1939), 'The Disorderly Motion in the *Timaeus*', *CQ* 33: 71–83.

——(1946), 'Solonian Justice', *CP* 41: 65–83.

——(1947), 'Equality and Justice in Early Greek Cosmologies', *CP* 42: 156–78.

——(1955), 'On Heraclitus', *AJP* 76: 337–68.

——(1965a), 'Creation in the *Timaeus*: Is it a Fiction?', in R. E. Allen (ed.), *Studies in Plato's Metaphysics* (London: Routledge & Kegan Paul), 401–19.

——(1965b), 'Minimal Parts in Epicurean Atomism', *Isis*, 56: 121–47.

——(1966), 'Zeno of Sidon as a Critic of Euclid', in L. Wallach (ed.), *The Classical Tradition* (Ithaca, NY: Cornell University Press), 148–59.

——(1967), 'Zeno of Elea', in Paul Edwards (ed.), *Encyclopedia of Philosophy* (New York: Macmillan), viii: 369–79.

——(1969), 'Reasons and Causes in the *Phaedo*', *PR* 78: 291–325.

VON ARNIM, HANS (1902), 'Die Weltperioden bei Empedocles', in *Festschrift Theodor Gomperz* (Vienna: A. Hoelder), 16–27.

*VON ARNIM, HANS (1931), *Die Enstehung der Gotteslehre des Aristoteles* (Vienna: Hölder-Pichler-Tempsky A.-G.).

*WAGNER, HANS (1967), *Aristoteles: Physikvorlesung* (Berlin: Akademie-Verlag).

WARDY, ROBERT (1990), *The Chain of Change: A Study of Aristotle's Physics VII* (Cambridge: Cambridge University Press).

WATERLOW, SARAH (1982), *Nature, Change, and Agency in Aristotle's Physics* (Oxford: Clarendon Press).

WHITE, MICHAEL J. (1992), *The Continuous and the Discrete* (Oxford: Clarendon Press).

*WICKSTEED, P. H., and *CORNFORD, F. M. (1929, 1934), *Aristotle: Physics*, 2 vols., Loeb Classical Library (Cambridge, Mass.: Harvard University Press).

WIELAND, WOLFGANG (1962), *Die aristotelische Physik* (Göttingen: Vandenhoek & Ruprecht).

WRIGHT, M. R. (1981), *Empedocles: The Extant Fragments* (New Haven: Yale University Press).

ZELLER, EDUARD (1876/1881), *A History of Greek Philosophy: From the Earliest Period to the Time of Socrates*, trans. S. F. Alleyne from the 2nd edn., 2 vols. (London: Longmans, Green, and Co.).

ZEYL, DONALD J. (1987), 'Commentary on Robinson', *PBACAP* 1: 120–5.

GLOSSARY

ἀΐδιος	*aïdios*	everlasting
αἰτία	*aitia*	cause, explanation
αἴτιον	*aition*	cause; responsible
ἄκρος	*akros*	extreme
ἀλλοίωσις; ἀλλοιοῦσθαι	*alloiōsis; alloiousthai*	alternation; alter
ἀνακάμπτειν	*anakamptein*	double back
ἀντικείμενον	*antikeimenon*	opposite
ἀντιπερίστασις	*antiperistasis*	recirculation
ἁπλός; ἁπλῶς	*haplos; haplōs*	simple; without qualification
ἀπογίγνεσθαι	*apogignesthai*	depart
ἀρχή	*archē*	principle, source, starting-point, beginning
αὔξησις; αὐξάνεσθαι	*auxēsis; auxanesthai*	(n.) increase; (vb.) increase
βίᾳ	*biāi*	by force
γένεσις; γίγνεσθαι	*genesis; gignesthai*	coming to be; come to be, arrive
δεικνύναι	*deiknunai*	prove; show
διάκρισις; διακρίνειν	*diakrisis; diakrinein*	segregation; segregate
δύναμις; δυνάμει	*dunamis; dunamei*	power, potentiality; potentially
εἶδος	*eidos*	form
ἐναντίον; ἐναντίωσις	*enantion; enantiōsis*	contrary; contrariety
ἐνέργεια	*energeia*	actuality
ἐντελέχεια	*entelecheia*	realization, actuality
ἐφεξῆς	*ephexēs*	successive, subsequent
ἐχόμενον	*echomenon*	contiguous
ἠρεμεῖν	*ēremein*	rest; be at rest
ἱστάναι	*histanai*	(tr.) stop; (intr.) halt, stand still, come to a standstill
καθ᾽ αὑτό	*kath' hauto*	intrinsic
κατὰ συμβεβηκός	*kata sumbebēkos*	incidental
κινεῖν	*kinein*	cause motion, move (tr.)
κινεῖσθαι	*kineisthai*	be moved, be in motion, move (intr.)
κίνησις	*kinēsis*	motion; movement

κινοῦν	*kinoun*	mover
μάνωσις	*manōsis*	rarefaction
μέγεθος	*megethos*	size, magnitude
μεταβολή; μεταβάλλειν	*metabolē; metaballein*	(n.) change; (vb.) change
μεταξύ	*metaxu*	intermediate, interval
μικτός	*miktos*	(of motions) combination
νῦν, τό	*nun, to*	the now
ὁμαλής	*homalēs*	uniform
ὄντα	*onta*	beings, things
περιφερής	*peripherēs*	circumference, arc
πύκνωσις	*puknōsis*	condensation
ῥιπτόμενον	*rhiptomenon*	projectile
στέρησις	*sterēsis*	privation
σύγκρισις; συγκρίνειν	*sunkrisis; sunkrinein*	aggregation; aggregate
συνάπτειν	*sunaptein*	be contiguous
συνεχής	*sunechēs*	continuous
ὑστερίζειν	*husterizein*	lag behind
φαντασία	*phantasia*	imagination
φέρεσθαι	*pheresthai*	travel, (tr.) traverse
φθίσις; φθίνειν	*phthisis; phthinein*	(n.) decrease; (vb.) decrease
φθορά; φθείρειν	*phthora; phtheirein*	perishing; perish
φορά	*phora*	locomotion, change of place

INDEX LOCORUM

GENERAL INDEX

motion (*cont.*):
  uniform 265^{b}11–14; 119–20
  see also change; locomotion
Mourelatos, A. P. D. 191, 192

natural cycles 49
natural place 130
Newton, I. 86, 95, 135, 142, 162, 173,
  175
now 251^{b}19–23
Nussbaum, M. C. 72

O'Brien, D. 40, 52
obstacle 255^{b}17–31
Ockham's razor 108
ontological principle 118
opinion 243^{a}25–30
opposites 261^{a}32–^{b}26
organs 81
Osborne, C. 40
Owen, G. E. L. 57, 59, 72

Parmenides x–xi, 43–4, 63, 64–5, 118,
  150, 165
patient 41
Pavlov, I. P. 112–13
Penner, T. 59, 135, 192
perfect tense predication 139, 142,
  144–5
peripatetic tradition 178
phantasy 252^{a}5
Philoponus, J. 111, 123, 134, 153, 161,
  176
pineal gland 181
plants 259^{b}2–3
Plato xi–xii, xv, 37
  analysis of cause 90
  contrary properties 99
  forms xvi, 181
  generates time 251^{b}17–18
  immortal soul 50
  original motion self-motion 42
  projectile motion 176
  world as living creature 116–17
Pluralists x, xi, 38, 164–5
Plutarch 46, 56
points on a line 262^{a}20–8, 262^{b}5–7,
  137–8, 161
Popper, K. 66
Power:
  active vs. passive 76–7, 86, 89, 173–4
  distinction from nature 76–9, 86, 89
  greater 266^{b}7–8

infinite and finite 266^{a}24–^{b}24
in medium 267^{a}2–15
see also *dunamis*
potentiality and actuality 251^{a}8–17,
  255^{a}30–^{b}13, ^{b}17–29
  applied to perfect tense predication
    142
  of back and forth motion 150–1
  of circular and straight motion 162
  of essential cause 85
  of midpoint 262^{a}20–8
  priority of actuality 127
  relation to endpoint 142
  relation to original mover 257^{a}33–
    ^{b}13
  Simplicius' remarks on 88
primary 260^{b}16–17, 265^{a}22–7, ^{b}8–11,
  ^{b}23–9
Principle of Plenitude 44, 107
Principle of Sufficient Reason 54, 161
principle of uniformity of nature 56
process and event 191–2
Proclus 46, 57
projectile, *see* motion, projectile
properties 107–8
Protagoras 134
prothesis 152–3
Pythagoreans 57, 147

rarefaction, *see* condensation and
  rarefaction
rational vs. irrational potencies 44–5,
  79–80, 82
Raven, J. E. 147
recirculation 176
regularity theory of causation 56
Reinhardt, K. 66
rest 251^{a}17–28, 252^{b}12–16, 253^{a}7–21,
  261^{b}1–2, ^{b}3; 40–1
Robinson, T. M. 47
Ross, W. D. xiv, 39–40, 47, 50, 82, 95–
  6, 108, 110, 148, 155, 162
Ryle, G. 68

Same and Different, circles of 119
seed 70
self-mover 258^{a}27–^{b}4, 259^{b}20–2,
  261^{a}23–6; 89, 98–9
sense experience 253^{a}32–^{b}2; 72–3
set theory 144
Sextus Empiricus 147
Sicilian Muses 53
simple substance 81